D0915520

CAMBRIDGE TRACTS IN MATHEMATICS

General Editors



---

**168** **The Cube: A Window to Convex and Discrete Geometry**

# THE CUBE: A WINDOW TO CONVEX AND DISCRETE GEOMETRY

CHUANMING ZONG
*Peking University*

CAMBRIDGE UNIVERSITY PRESS
Cambridge, New York, Melbourne, Madrid, Cape Town, Singapore, São Paulo

CAMBRIDGE UNIVERSITY PRESS
The Edinburgh Building, Cambridge CB2 2RU, UK

Published in the United States of America by Cambridge University Press, New York

www.cambridge.org
Information on this title: www.cambridge.org/9780521855358



First published 2006

Printed in the United Kingdom at the University Press, Cambridge

*A catalog record for this book is available from the British Library*

ISBN-13 978-0-521-85535-8 hardback
ISBN-10 0-521-85535-7 hardback

# Contents

# Preface

*What is the simplest object in an n-dimensional Euclidean space?* The answer should be a unit cube or a unit ball. On the one hand, compared with others, they can be easily described and intuitively imagined; on the other hand, they are perfect in shape with respect to symmetry and regularity. However, in fact, neither of them is really simple. For example, to determine the density of the densest ball packing is one of the most challenging problems in mathematics. As for the unit cube, though no open problem is as famous as Kepler's conjecture, there are many fascinating problems of no less importance.

*What is the most important object in the n-dimensional Euclidean space?* The answer should be the unit cube and the unit ball as well. Since, among other things, the unit ball is key to understanding metrics and surface area, and the unit cube is key to understanding measure and volume, in addition, a high-dimensional unit cube is rich in structure and geometry. Deep understanding of the unit cube is essential to understanding combinatorics and $n$-dimensional geometry.

This book has two main purposes: to show what is known about the cube and to demonstrate how analysis, combinatorics, hyperbolic geometry, number theory, algebra, etc. can be applied to the study of the cube.

The first two chapters discuss the area of a cross section and the area of a projection of the cube. The key problems in these directions are: *What is the maximal (minimal) area of a k-dimensional cross section of an n-dimensional unit cube? What is the maximal (minimal) area of a k-dimensional projection of an n-dimensional unit cube?* These problems are very natural and simple sounding. However, their solutions have not been completely discovered yet! On the other hand, it will be shown how deep analysis and linear algebra can deal with problems of these types.

The next two chapters study the maximal simplex inscribed in the cube and the triangulation problem. For example: *What is the maximum volume of a k-dimensional simplex that can be inscribed into an n-dimensional unit cube? What is the smallest number of simplices needed to triangulate an n-dimensional cube?* These problems are very combinatorial in nature. The inscribed simplex problem in fact is closely related to Hadamard matrices. However, to have a deep understanding of these problems, we have to apply knowledge of number theory and even hyperbolic geometry! In addition, so far our knowledge about these problems is very limited.

Then we discuss the 0/1 polytopes. Clearly 0/1 polytopes do provide nice models for finite geometry, geometric combinatorics, and optimization. However, what we want to emphasize here is the connection with coding theory. In fact, many problems about binary codes are problems about 0/1 polytopes!

The last three chapters deal with Minkowski's conjecture, Furtwängler's conjecture, and Keller's conjecture, respectively. The conjectures themselves are fascinating; moreover their algebraic solutions are just magic! According to S. K. Stein, "which in total are almost as startling as the metamorphosis of a caterpillar to a butterfly." Is there any other fascinating mathematical problem solved in a more surprising way? I know not one.

This work has been supported by the National Science Foundation of China, 973 Project, and a special grant from Peking University. Parts of it were written while I was a visitor at the Institut des Hautes Études Scientifiques and Mathematical Sciences Research Institute. I am very grateful to these institutions for their support. For the helpful comments, remarks, and suggestions, I am very grateful to R. Astley, I. Bárány, R. J. Gardner, P. M. Gruber, M. Henk, J. Liu, W. Orrick, B. Solomon, J. M. Wills, L. Yu, T. Zhou, and G. M. Ziegler.

# Basic notation

| | |
|---|---|
| $E^n$ | Euclidean space of $n$-dimensions. |
| $\mathbf{x}$ | A point (or a vector) of $E^n$ with coordinates $(x_1, x_2, \ldots, x_n)$; or an element of a group. |
| $\mathbf{o}$ | The origin of $E^n$. |
| $\langle \mathbf{x}, \mathbf{y} \rangle$ | The inner product of two vectors $\mathbf{x}$ and $\mathbf{y}$. |
| $\|\mathbf{x}, \mathbf{y}\|$ | The Euclidean distance between two points $\mathbf{x}$ and $\mathbf{y}$. |
| $\|\mathbf{x}\|$ | The Euclidean norm of $\mathbf{x}$. |
| $X$ | A set of points in $E^n$. |
| $\mathrm{conv}\{X\}$ | The convex hull of $X$. |
| $\mathrm{int}(X)$ | The interior of $X$. |
| $v(X)$ | The volume of $X$. |
| $s(X)$ | The surface area of $X$. |
| $v_k(X)$ | The $k$-dimensional measure of $X$. |
| $K$ | An $n$-dimensional convex body. |
| $W_i(K)$ | The $i$th quermassintegral of $K$. |
| $I^n$ | The $n$-dimensional unit cube $\{\mathbf{x} : \; \lvert x_i \rvert \leq \frac{1}{2}\}$. |
| $\overline{I^n}$ | The $n$-dimensional unit cube $\{\mathbf{x} : \; 0 \leq x_i \leq 1\}$. |
| $B^n$ | The $n$-dimensional unit ball centered at $\mathbf{o}$. |
| $\omega_n$ | The volume of an $n$-dimensional unit ball. |
| $P$ | An $n$-dimensional polytope. |
| $V(P)$ | The set of the vertices of $P$. |
| $Z$ | The set of all integers. |
| $R$ | The set of all real numbers. |
| $\Lambda$ | An $n$-dimensional lattice. |
| $Z^n$ | The $n$-dimensional lattice $\{(z_1, z_2, \ldots, z_n) : z_i \in Z\}$. |
| $\alpha(n, k)$ | The maximal volume of a $k$-dimensional cross section of $I^n$. |
| $\beta(n, k)$ | The maximal volume of a $k$-dimensional projection of $I^n$. |

| | |
|---|---|
| $\theta(n,k)$ | The maximal value of $\det(AA')$, where $A$ is a $k \times n$ binary matrix. Especially, we abbreviate $\theta(n,n)$ to $\theta_n$. |
| $\theta^*(n,k)$ | The maximal value of $\det(AA')$, where $A$ is a $k \times n$ matrix with $\pm 1$ elements. Especially, we abbreviate $\theta^*(n,n)$ to $\theta_n^*$. |
| $\gamma(n,k)$ | The maximal volume of a $k$-dimensional simplex inscribed in $I^n$. Especially, we abbreviate $\gamma(n,n)$ to $\gamma_n$. |
| $\tau_n$ | The minimal cardinality of a triangulation of $I^n$. |
| $A(n,s)$ | The maximal cardinality of an $n$-dimensional binary code with separation $s$. |
| $\phi(n)$ | The number of the $n$-dimensional 0/1 polytopes reduced from $\overline{I^n}$. |
| $\langle \mathbf{g} \rangle$ | The cyclic group generated by $\mathbf{g}$. |
| $\lvert \mathbf{g} \rvert$ | The order of $\mathbf{g}$. |
| $Z_k$ | The additive cyclic group $\{0, 1, \ldots, k-1\}$ modulo $k$. |
| $O_k$ | The multiplicative cyclic group $\left\{1, e^{2\pi i/k}, \ldots, e^{2(k-1)\pi i/k}\right\}$, where $i = \sqrt{-1}$. |
| $\Re(G)$ | The group ring generated by a group $G$. |

# Introduction

Let $E^n$ denote the $n$-dimensional *Euclidean space* over the real number field $R$ and with an *orthonormal basis* $\{\mathbf{e}_1, \mathbf{e}_2, \ldots, \mathbf{e}_n\}$. Then each point $\mathbf{x}$ of $E^n$ can be uniquely expressed as

$$\mathbf{x} = x_1\mathbf{e}_1 + x_2\mathbf{e}_2 + \cdots + x_n\mathbf{e}_n = (x_1, x_2, \ldots, x_n),$$

where $x_i$ is known as the $i$th *coordinate* of $\mathbf{x}$ with respect to the basis. Let $\langle \mathbf{u}, \mathbf{v} \rangle$ denote the *inner product* of two vectors $\mathbf{u}$ and $\mathbf{v}$ and let $\|\mathbf{x}, \mathbf{y}\|$ denote the *Euclidean distance* between two points $\mathbf{x}$ and $\mathbf{y}$, by which the *Euclidean metric* is defined. With the coordinates of the vectors and the points, we can write

$$\langle \mathbf{u}, \mathbf{v} \rangle = \sum_{i=1}^{n} u_i v_i$$

and

$$\|\mathbf{x}, \mathbf{y}\| = \left( \sum_{i=1}^{n} (x_i - y_i)^2 \right)^{1/2}.$$

For convenience, we abbreviate $\|\mathbf{x}, \mathbf{o}\|$ to $\|\mathbf{x}\|$. Let $\theta$ denote the angle between $\mathbf{u}$ and $\mathbf{v}$, then we have

$$\langle \mathbf{u}, \mathbf{v} \rangle = \|\mathbf{u}\| \cdot \|\mathbf{v}\| \cdot \cos\theta.$$

Clearly two vectors are orthogonal to each other if and only if their inner product is zero.

Now let us introduce two particular objects in $E^n$, namely

$$I^n = \left\{ \mathbf{x} \in E^n : \ |x_i| \le \tfrac{1}{2} \text{ for all } i \right\}$$

and

$$B^n = \left\{ \mathbf{x} \in E^n : \ \|\mathbf{x}\| \le 1 \right\}.$$

A subset of $E^n$ is called an *n-dimensional unit cube* if it is congruent to $I^n$, and is called an *n-dimensional unit ball* if it is congruent to $B^n$. For convenience of the forthcoming usage, we introduce another particular unit cube

$$\overline{I^n} = \left\{ \mathbf{x} \in E^n : \ 0 \leq x_i \leq 1 \text{ for all } i \right\}.$$

In fact, $\overline{I^n}$ is a translate of $I^n$ with a translative vector $\mathbf{v} = (\frac{1}{2}, \frac{1}{2}, \ldots, \frac{1}{2})$.

In the $n$-dimensional Euclidean space, the *volume* $v(X)$ of a set $X$ is its *Lebesgue measure*; that is

$$v(X) = \int_{E^n} \chi(\mathbf{x})\, d\mathbf{x},$$

where $\chi(\mathbf{x})$ is the *characteristic function* of the set. For the unit cube $I^n$ and the unit ball $B^n$, we have

$$v(I^n) = 1 \tag{0.1}$$

and

$$v(B^n) = \frac{\pi^{\frac{n}{2}}}{\Gamma(\frac{n}{2}+1)}, \tag{0.2}$$

where $\Gamma(x)$ is the *gamma function*. For convenience, we will abbreviate the volume of the $n$-dimensional unit ball to $\omega_n$. In fact, (0.1) is the foundation in defining the measure of a general set in $E^n$.

For two subsets $C$ and $D$ of $E^n$, we define their *Minkowski sum* as

$$C + D = \left\{ \mathbf{x} + \mathbf{y} : \ \mathbf{x} \in C, \ \mathbf{y} \in D \right\}.$$

Then the *surface area* $s(X)$ of the set $X$ is defined by

$$s(X) = \lim_{\epsilon \to 0} \frac{v(X + \epsilon B^n) - v(X)}{\epsilon},$$

if the limit does exist. Intuitively speaking, the set $X + \epsilon B^n$ is nothing else but the result of putting a tight coat of thickness $\epsilon$ on $X$. Applying this formula to $I^n$ and $B^n$, respectively, we can easily deduce

$$s(I^n) = 2n$$

and

$$s(B^n) = n \cdot \omega_n.$$

A subset $K$ of $E^n$ is *convex* if

$$\lambda \mathbf{x} + (1 - \lambda)\mathbf{y} \in K,$$

whenever both $\mathbf{x}$ and $\mathbf{y}$ belong to $K$ and $0 < \lambda < 1$. In addition, if it is also compact, we call it a *convex body*. For example, all balls, cubes, and simplices are convex bodies. Let $\mathbf{x}$ and $\mathbf{y}$ be two points of the unit ball $B^n$ and let $\lambda$ be a number satisfying $0 < \lambda < 1$. Then, by the *Cauchy–Schwarz inequality* we get

$$\begin{aligned}\|\lambda\mathbf{x}+(1-\lambda)\mathbf{y}\| &= \left(\sum_{i=1}^{n}(\lambda x_i+(1-\lambda)y_i)^2\right)^{1/2}\\ &\leq \lambda\left(\sum_{i=1}^{n}x_i^2\right)^{1/2}+(1-\lambda)\left(\sum_{i=1}^{n}y_i^2\right)^{1/2}\\ &\leq 1.\end{aligned}$$

Therefore, the unit ball is indeed convex. The convexity of the cubes and the simplices can be deduced by similar routine arguments.

There is another important concept about convexity, which will be useful in this book; namely, the *convex hull* of a given set $X$, which is defined by

$$\mathrm{conv}\{X\} = \left\{\sum \lambda_i\mathbf{x}_i : \mathbf{x}_i \in X;\ \lambda_i \geq 0,\ \sum \lambda_i = 1\right\}.$$

In fact, by *Carathéodory's theorem*, we can restrict each of the sum over only $n+1$ terms. In particular, if $\mathrm{card}\{X\}$ is a finite number, then $\mathrm{conv}\{X\}$ is a *convex polytope*. For example, both a cube and a simplex are polytopes and they are the convex hulls of the sets of their vertices. Let $H$ be a supporting hyperplane of an $n$-dimensional polytope $P$. We call $F = P \cap H$ a *k-face* of $P$ if it is $k$-dimensional. In particular, an $(n-1)$-face is called a *facet* and a 0-face is a vertex.

Let $Z$ denote the *ring of integers* and let $\mathbf{a}_1,\ \mathbf{a}_2, \ldots, \mathbf{a}_n$ be $n$ linearly independent vectors in $E^n$, then the set

$$\Lambda = \left\{\sum_{i=1}^{n} z_i\mathbf{a}_i : z_i \in Z\right\}$$

is called an $n$-dimensional *lattice* and $\{\mathbf{a}_1, \mathbf{a}_2, \ldots, \mathbf{a}_n\}$ is called a *basis* of the lattice. It is easy to see that a lattice is a *free abelian group* under addition and there are many different bases for a given lattice. For example

$$Z^n = \left\{(z_1, z_2, \ldots, z_n) : z_i \in Z\right\}$$

is an $n$-dimensional lattice with a basis $\{\mathbf{e}_1, \mathbf{e}_2, \ldots, \mathbf{e}_n\}$. In addition, the set $\{\mathbf{u}_1, \mathbf{u}_2, \ldots, \mathbf{u}_n\}$ is a basis for $Z^n$ if and only if $(u_{ij})$ is a unimodular integral matrix, where $\mathbf{u}_i = (u_{i1}, u_{i2}, \ldots, u_{in})$.

To end this brief introduction, let us have a close look at the unit cube. Clearly, an $n$-dimensional unit cube is a cylinder of height 1 over an $(n-1)$-dimensional one. In other words

$$I^n = I^1 \oplus I^{n-1}. \tag{0.3}$$

Therefore, it is easy to see that every $k$-face $(0 \le k \le n-1)$ of $I^n$ is a $k$-dimensional unit cube. Let $f(n,k)$ denote the number of the $k$-faces of $I^n$. By (0.3) we get

$$f(n,k) = 2f(n-1,k) + f(n-1,k-1). \tag{0.4}$$

Then, by induction on $n$ and the identity

$$\binom{n-1}{k} + \binom{n-1}{k-1} = \binom{n}{k},$$

it can be deduced from (0.4) that

$$f(n,k) = 2^{n-k}\binom{n}{k}.$$

As a conclusion, when $0 \le k \le n-1$, the unit cube $I^n$ has exactly $2^{n-k}\binom{n}{k}$ $k$-faces, each of which is a $k$-dimensional unit cube.

# 1
# Cross sections

## 1.1 Introduction

Let us start with the two-dimensional case. Let $H^1$ denote a straight line in $E^2$ passing through the origin $\mathbf{o}$ (a one-dimensional subspace of $E^2$) and let $\ell(I^2 \bigcap H^1)$ denote the length of $I^2 \bigcap H^1$. If, without loss of generality, $H^1$ intersects the boundary of $I^2$ at $(\frac{1}{2}, y)$ and $(-\frac{1}{2}, -y)$, then $|y| \leq \frac{1}{2}$ and

$$\ell(I^2 \bigcap H^1) = \sqrt{1+4y^2}.$$

Therefore, for every $H^1$, we have

$$1 \leq \ell(I^2 \bigcap H^1) \leq \sqrt{2}, \tag{1.1}$$

where the lower bound is attained if and only if $H^1$ is an axis of $E^2$ and the upper bound is attained if and only if $H^1$ contains a pair of antipodal vertices of $I^2$.

In fact, the length of any segment contained in $I^2$ is at most $\sqrt{2}$. This statement can be deduced by the following argument. Let $L$ be a segment contained in $I^2$. Since $I^2$ is centrally symmetric, the segment $L'$, which is symmetric to $L$, is also contained in $I^2$. Then by convexity we get

$$\tfrac{1}{2}(L+L') \subset I^2.$$

Since $L$ is parallel with $L'$ and $\frac{1}{2}(L+L')$ contains the origin $\mathbf{o}$, by (1.1) we get

$$\ell(L) = \ell(\tfrac{1}{2}(L+L')) \leq \sqrt{2}.$$

The three-dimensional case is more complicated and more interesting. First, we study the one-dimensional sections. Without loss of generality, we assume that $H^1$ intersects the boundary of $I^3$ at $(\frac{1}{2}, y, z)$ and $(-\frac{1}{2}, -y, -z)$. Then we have $|y| \leq \frac{1}{2}$, $|z| \leq \frac{1}{2}$, and

$$\ell(I^3 \bigcap H^1) = \sqrt{1+4y^2+4z^2}.$$

Therefore, for every $H^1$, we get

$$1 \le \ell(I^3 \cap H^1) \le \sqrt{3}, \tag{1.2}$$

where the lower bound is attained if and only if $H^1$ is an axis of $E^3$ and the upper bound is attained if and only if $H^1$ contains a pair of antipodal vertices of $I^3$. As with the two-dimensional case, by symmetry and convexity we can deduce that the length of any segment contained in $I^3$ is at most $\sqrt{3}$.

Next, let us discuss the two-dimensional cross sections of $I^3$. Let $\mathbf{u}$ be a point on the boundary of $I^3$ and let $H^2$ denote the two-dimensional subspace $\{\mathbf{x}:\ \mathbf{x} \in E^3,\ \langle \mathbf{x}, \mathbf{u}\rangle = 0\}$. Since $I^3$ has six facets and every edge of $I^3 \cap H^2$ is an intersection of $H^2$ with one of the facets, by symmetry it follows that $I^3 \cap H^2$ is either a parallelogram or a hexagon.

Let $v_2(I^3 \cap H^2)$ denote the area of $I^3 \cap H^2$ and write

$$U_1 = \left\{\mathbf{u}:\ u_3 = \tfrac{1}{2},\ u_1 \ge 0,\ u_2 \ge 0,\ u_1 + u_2 \le \tfrac{1}{2}\right\},$$

$$U_2 = \left\{\mathbf{u}:\ u_3 = \tfrac{1}{2},\ u_1 \le \tfrac{1}{2},\ u_2 \le \tfrac{1}{2},\ u_1 + u_2 > \tfrac{1}{2}\right\},$$

and, for $i = 1, 2, 3$

$$F_i = \left\{\mathbf{x} \in E^3:\ x_i = \tfrac{1}{2},\ |x_j| \le \tfrac{1}{2} \text{ for } j \ne i\right\}.$$

Now, we estimate $v_2(I^3 \cap H^2)$ by considering two cases.

**Case 1.** $\mathbf{u} \in U_1$. Then the corresponding plane $H^2$ does not intersect the relative interior of $F_3$ and therefore $I^3 \cap H^2$ is a parallelogram (see Figure 1.1). By projecting $I^3 \cap H^2$ on to $F_3$ we get

$$v_2(I^3 \cap H^2) = \frac{\sqrt{0.5^2 + u_1^2 + u_2^2}}{0.5} v_2(F_3).$$

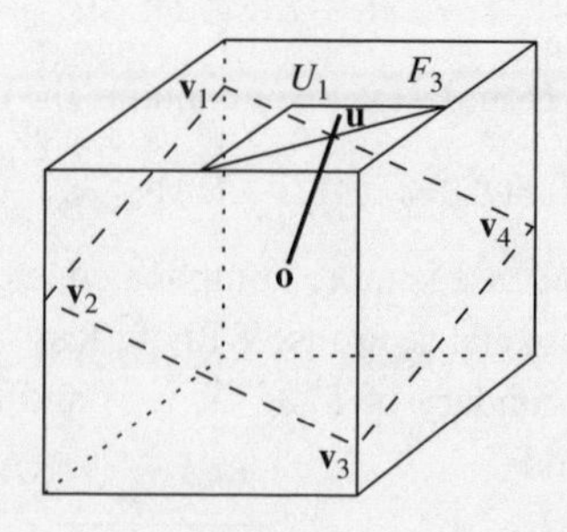


Figure 1.1

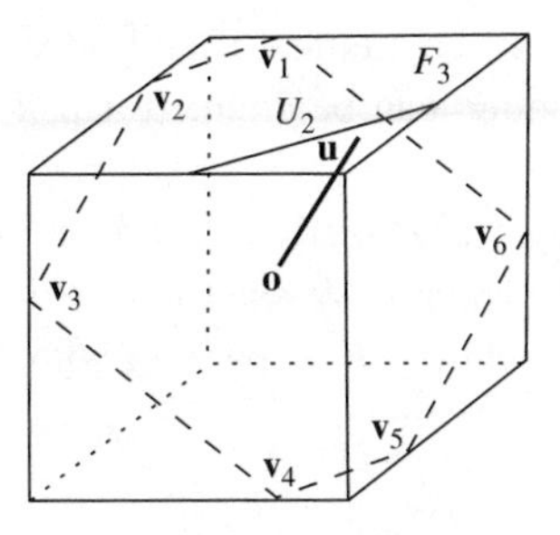


Figure 1.2

Thus, by a routine argument we can deduce

$$1 \leq v_2(I^3 \cap H^2) \leq \sqrt{2},$$

where the lower bound is attained if and only if $\mathbf{u} = (0, 0, \frac{1}{2})$ and the upper bound is attained if and only if $\mathbf{u} = (\frac{1}{2}, 0, \frac{1}{2})$ or $\mathbf{u} = (0, \frac{1}{2}, \frac{1}{2})$.

**Case 2.** $\mathbf{u} \in U_2$. Then the corresponding plane $H^2$ does intersect the relative interior of every facet of $I^3$ and therefore $I^3 \cap H^2$ is a hexagon (see Figure 1.2). Assume that $H^2$ does intersect the boundary of $F_3$ at two points $\mathbf{v}_1$ and $\mathbf{v}_2$. By a routine computation we can determine that $\mathbf{v}_1 = (\frac{2u_2-1}{4u_1}, -\frac{1}{2}, \frac{1}{2})$ and $\mathbf{v}_2 = (-\frac{1}{2}, \frac{2u_1-1}{4u_2}, \frac{1}{2})$. Since the projection of $I^3 \cap H^2$ to $F_3$ has area $1 - (\frac{1}{2} + \frac{2u_2-1}{4u_1})$ $(\frac{1}{2} + \frac{2u_1-1}{4u_2})$, we get

$$v_2(I^3 \cap H^2) =$$
$$\sqrt{1 + (2u_1)^2 + (2u_2)^2} \left[ 1 - \left( \frac{1}{2} + \frac{2u_2 - 1}{4u_1} \right) \left( \frac{1}{2} + \frac{2u_1 - 1}{4u_2} \right) \right].$$

For any fixed number $c$, it is easy to see that

$$1 - \left( \frac{1}{2} + \frac{2u_2 - 1}{4u_1} \right) \left( \frac{1}{2} + \frac{2u_1 - 1}{4u_2} \right) = c$$

is a quadratic curve which is symmetric with respect to the straight line $u_1 = u_2$. Therefore, in this case $v_2(I^3 \cap H^2)$ attains its minimum and maximum on the boundary of $U_2$ or on the line $u_1 = u_2$. By checking these subcases we get

$$\sqrt{\frac{3}{2}} < v_2(I^3 \cap H^2) < \sqrt{2}.$$

As a conclusion, by symmetry, for every $H^2$ we have

$$1 \leq v_2(I^3 \cap H^2) \leq \sqrt{2}, \tag{1.3}$$

where the lower bound can be attained if and only if $\mathbf{u}$ is in the direction of an axis and the upper bound can be attained if and only if $H^2$ contains two pairs of antipodal vertices of $I^3$.

In fact, the area of any planar section of $I^3$ is at most $\sqrt{2}$. Let $P$ be such a section. Then the set $P'$, which is symmetric to $P$ with respect to $\mathbf{o}$, is also a planar section of $I^3$. In addition, $P'$ is parallel with $P$. Therefore, we have

$$\mathbf{o} \in \tfrac{1}{2}(P+P') \subset I^3$$

and, by the *Brunn–Minkowski inequality* and (1.3)

$$v_2(P) \leq v_2(\tfrac{1}{2}(P+P')) \leq \sqrt{2}.$$

These examples are relatively simple, at least they are manageable by elementary methods. However, similar problems in higher dimensions are much more challenging and fascinating. Let $H^k$ denote a $k$-dimensional linear subspace of $E^n$ containing $\mathbf{o}$ and let $v_k(S)$ denote the $k$-dimensional volume (measure) of a set $S$. The purpose of this chapter is to study the measure and the structure of $I^n \bigcap H^k$.

## 1.2 Good's conjecture

Based on the examples listed in the previous section, according to Hensley (1979), Anton Good made the following conjecture about the lower bound for $v_k(I^n \bigcap H^k)$.

**Good's conjecture.** *Let $k$ be an integer satisfying $1 \leq k \leq n-1$. For every $k$-dimensional subspace $H^k$ of $E^n$, we have*

$$v_k(I^n \bigcap H^k) \geq 1.$$

This conjecture is simple-sounding in nature. However, we have to use complicated analytic machinery to prove it. Let $\overline{B^k}$ denote the $k$-dimensional ball with radius

$$r = \frac{\Gamma(\frac{k}{2}+1)^{\frac{1}{k}}}{\pi^{\frac{1}{2}}}$$

and centered at the origin of $E^k$. Then by (0.2) we have

$$v_k(\overline{B^k}) = \omega_k \cdot r^k = \frac{\pi^{\frac{k}{2}}}{\Gamma(\frac{k}{2}+1)} \cdot \frac{\Gamma(\frac{k}{2}+1)}{\pi^{\frac{k}{2}}} = 1.$$

Let $\chi(S, \mathbf{x})$ denote the characteristic function of a subset $S$ of $E^n$. In 1979 J. D. Vaaler proved the following theorem.

**Theorem 1.1 (Vaaler, 1979).** *Suppose that* $n_1, n_2, \ldots, n_j$ *are positive integers,* $n = n_1 + n_2 + \cdots + n_j$, $D = \overline{B^{n_1}} \oplus \overline{B^{n_2}} \oplus \cdots \oplus \overline{B^{n_j}} \subset E^n$, *and* $A$ *is a real* $k \times n$ *matrix with rank* $k$. *Then we have*

$$\int_{E^k} \chi(D, \mathbf{x}A)\, d\mathbf{x} \geq |AA'|^{-\frac{1}{2}}, \tag{1.4}$$

*where* $A'$ *is the transpose of* $A$.

As usual, $X \oplus Y$ means the *Cartesian product* of $X$ and $Y$. Let us take $n_1 = n_2 = \cdots = n_j = 1$ (then $j = n$) and choose $A$ in this theorem so that its rows form an orthonormal basis for $H^k$ in $E^n$. Then $D$ is nothing else but the unit cube $I^n$ and

$$\int_{E^k} \chi(I^n, \mathbf{x}A)\, d\mathbf{x} = \int_{H^k} \chi(I^n, \mathbf{y})\, d\mathbf{y} = v_k(I^n \cap H^k).$$

On the other hand, by the assumption, we get

$$|AA'| = 1.$$

Therefore, Good's conjecture follows as a corollary of Theorem 1.1.

**Corollary 1.1 (Vaaler, 1979).** *For every* $k$*-dimensional subspace* $H^k$ *of* $E^n$, *we have*

$$v_k(I^n \cap H^k) \geq 1.$$

For a deep generalization of this result we refer to Meyer and Pajor (1988). Vaaler's theorem is very geometric. However, as one can imagine, its proof is very analytical. To introduce the proof, let us start with a couple of definitions.

**Definition 1.1.** *Let* $f(\mathbf{x})$ *be a nonnegative function defined in* $E^n$. *If*

$$f(\lambda \mathbf{x}_1 + (1-\lambda)\mathbf{x}_2) \geq f(\mathbf{x}_1)^{\lambda} f(\mathbf{x}_2)^{1-\lambda}$$

*holds for every pair of points* $\mathbf{x}_1$ *and* $\mathbf{x}_2$ *in* $E^n$ *and every* $\lambda$ *satisfying* $0 < \lambda < 1$, *then* $f(\mathbf{x})$ *is said to be logconcave.*

Let $g(\mathbf{x})$ be a *concave function* defined in $E^n$ and define $f(\mathbf{x}) = e^{g(\mathbf{x})}$. Since

$$g(\lambda \mathbf{x}_1 + (1-\lambda)\mathbf{x}_2) \geq \lambda g(\mathbf{x}_1) + (1-\lambda) g(\mathbf{x}_2)$$

holds for every pair of points $\mathbf{x}_1$ and $\mathbf{x}_2$ in $E^n$ and every $\lambda$ with $0 < \lambda < 1$, we have

$$\begin{aligned} f(\lambda \mathbf{x}_1 + (1-\lambda)\mathbf{x}_2) &= e^{g(\lambda \mathbf{x}_1 + (1-\lambda)\mathbf{x}_2)} \\ &\geq e^{\lambda g(\mathbf{x}_1)} \cdot e^{(1-\lambda) g(\mathbf{x}_2)} \\ &= f(\mathbf{x}_1)^{\lambda} f(\mathbf{x}_2)^{1-\lambda}. \end{aligned}$$

Thus $f(\mathbf{x})$ is logconcave in $E^n$.

**Definition 1.2.** *Let $\mu$ be a probability measure on $E^n$. If*

$$\mu(\lambda K_1 + (1-\lambda)K_2) \geq \mu(K_1)^\lambda \mu(K_2)^{1-\lambda}$$

*holds for every pair of open convex sets $K_1$ and $K_2$ in $E^n$ and every $\lambda$ with $0 < \lambda < 1$, then $\mu$ is said to be logconcave in $E^n$.*

As usual, we define the *support* of a probability measure $\mu$ to be the smallest closed subset $S$ of $E^n$ such that $\mu(E^n \setminus S) = 0$. In other words, the support of $\mu$ is the set of all points $\mathbf{x} \in E^n$ such that

$$\mu(rB^n + \mathbf{x}) > 0$$

for all $r > 0$. For convenience, it is denoted by $\mathrm{supp}(\mu)$.

Comparing with the logconcave functions, it seems more difficult to get examples for logconcave probability measures. In fact, as we can see from the following lemma, they are closely related.

**Lemma 1.1 (Borell, 1975 and Prékopa, 1973).** *Let $\mu$ be a logconcave probability measure on $E^n$ and suppose that* $\mathrm{supp}(\mu)$ *spans a k-dimensional subspace $H^k$ of $E^n$. Then there is a logconcave probability density function $f(\mathbf{x})$ defined on $H^k$ such that $d\mu = f dv_k$, where $v_k$ is the k-dimensional Lebesgue measure on $H^k$. On the other hand, for any logconcave probability density function $f(\mathbf{x})$ defined on a k-dimensional subspace $H^k$ in $E^n$, the probability measure defined by $d\mu = f dv_k$ is logconcave.*

The first part of this lemma was proved by C. Borell, and the second part was proved by A. Prékopa. Both this result and the next lemma are intuitively imaginable. However, like many results in measure theory, their proofs are very complicated.

**Proof (A sketch).** Let $\mathbf{x}$ and $\mathbf{y}$ be two distinct points in $H^k$ and let $\lambda$ be a number satisfying $0 < \lambda < 1$. Since $\mu$ is a logconcave probability measure with a density function $f(\mathbf{x})$

$$\mu\left(\epsilon B^k + \lambda\mathbf{x} + (1-\lambda)\mathbf{y}\right) \geq \mu\left(\epsilon B^k + \mathbf{x}\right)^\lambda \mu\left(\epsilon B^k + \mathbf{y}\right)^{1-\lambda}$$

holds for all sufficiently small $\epsilon$. Therefore, we can deduce

$$f(\lambda\mathbf{x} + (1-\lambda)\mathbf{y}) \geq f(\mathbf{x})^\lambda f(\mathbf{y})^{1-\lambda},$$

which means that $f(\mathbf{x})$ is logconcave.

To prove the second part, we start with a basic inequality (see Prékopa, 1971 for the details). *If $g_1(x)$, $g_2(x)$, and $g(x)$ are nonnegative Borel measurable functions satisfying*

$$g(z) = \sup_{x+y=2z} g_1(x)\, g_2(y),$$

*then we have*

$$\int_{-\infty}^{\infty} g(z)\,dz \ge \left(\int_{-\infty}^{\infty} g_1(x)^2 dx\right)^{\frac{1}{2}} \left(\int_{-\infty}^{\infty} g_2(y)^2 dy\right)^{\frac{1}{2}}.$$

Applying this inequality to the multivariable case

$$g(\mathbf{z}) = \sup_{\mathbf{x}+\mathbf{y}=2\mathbf{z}} g_1(\mathbf{x})\, g_2(\mathbf{y}),$$

we get

$$\int_{-\infty}^{\infty} g(\mathbf{z})\,dz_1 \ge \sup_{\substack{x_2+y_2=2z_2\\ \cdots \\ x_n+y_n=2z_n}} \left(\int_{-\infty}^{\infty} g_1(\mathbf{x})^2 dx_1\right)^{\frac{1}{2}} \left(\int_{-\infty}^{\infty} g_2(\mathbf{y})^2 dy_1\right)^{\frac{1}{2}}$$

and, by induction

$$\int_{E^n} g(\mathbf{z})\,d\mathbf{z} \ge \left(\int_{E^n} g_1(\mathbf{x})^2 d\mathbf{x}\right)^{\frac{1}{2}} \left(\int_{E^n} g_2(\mathbf{y})^2 d\mathbf{y}\right)^{\frac{1}{2}}. \tag{1.5}$$

If both $g_1(\mathbf{x})$ and $g_2(\mathbf{x})$ are logconcave, $m = 2^l$, where $l$ is a positive integer, $\frac{i}{m} = \lambda$, $\frac{j}{m} = 1-\lambda$, and

$$g(\mathbf{z}) = \sup_{\lambda\mathbf{x}+(1-\lambda)\mathbf{y}=\mathbf{z}} g_1(\mathbf{x})\, g_2(\mathbf{y}).$$

By the logconcavity and subsequent applications of (1.5), we get[1]

$$\begin{aligned}\int_{E^n} g(\mathbf{z})\,d\mathbf{z} &= \int_{E^n} \sup_{\lambda\mathbf{x}+(1-\lambda)\mathbf{y}=\mathbf{z}} g_1(\mathbf{x})\, g_2(\mathbf{y})\,d\mathbf{z}\\ &\ge \int_{E^n} \sup_{\sum \frac{\mathbf{x}_w}{m}=\mathbf{z}} \left\{\prod_{w=1}^{i} g_1(\mathbf{x}_w)^{\frac{1}{i}} \prod_{w=i+1}^{m} g_2(\mathbf{x}_w)^{\frac{1}{j}}\right\} d\mathbf{z}\\ &\ge \left(\int_{E^n} g_1(\mathbf{x})^{\frac{1}{\lambda}} d\mathbf{x}\right)^{\lambda} \left(\int_{E^n} g_2(\mathbf{x})^{\frac{1}{1-\lambda}} d\mathbf{x}\right)^{1-\lambda}. \end{aligned} \tag{1.6}$$

Finally, we define

$$g_1(\mathbf{x}) = \begin{cases} f(\mathbf{x}) & \text{if } \mathbf{x}\in K_1,\\ 0 & \text{otherwise,}\end{cases}$$

$$g_2(\mathbf{x}) = \begin{cases} f(\mathbf{x}) & \text{if } \mathbf{x}\in K_2,\\ 0 & \text{otherwise,}\end{cases}$$

[1] Prékopa's paper has an error in the corresponding formula of the last step of (1.6).

and

$$g(\mathbf{x}) = \begin{cases} f(\mathbf{x}) & \text{if } \mathbf{x} \in \lambda K_1 + (1-\lambda)K_2, \\ 0 & \text{otherwise.} \end{cases}$$

Then, by the logconcavity and (1.6), we get

$$\begin{aligned} \int_{\lambda K_1+(1-\lambda)K_2} f(\mathbf{x})\, d\mathbf{x} &= \int_{E^n} g(\mathbf{x})\, d\mathbf{x} \\ &\geq \int_{E^n} \sup_{\lambda \mathbf{x}+(1-\lambda)\mathbf{y}=\mathbf{z}} g_1(\mathbf{x})^{\lambda} g_2(\mathbf{y})^{1-\lambda} d\mathbf{z} \\ &\geq \left(\int_{E^n} g_1(\mathbf{x})\, d\mathbf{x}\right)^{\lambda} \left(\int_{E^n} g_2(\mathbf{x})\, d\mathbf{x}\right)^{1-\lambda} \\ &= \left(\int_{K_1} f(\mathbf{x})\, d\mathbf{x}\right)^{\lambda} \left(\int_{K_2} f(\mathbf{x})\, d\mathbf{x}\right)^{1-\lambda}. \end{aligned}$$

Therefore, $\mu$ is logconcave. □

**Definition 1.3.** *Let $\mu_1$ and $\mu_2$ be two probability measures on $E^n$. If*

$$\mu_1(C) \geq \mu_2(C)$$

*holds for every centrally symmetric convex body $C$, then we say that $\mu_1$ is more peaked than $\mu_2$. Let $f_1(\mathbf{x})$ and $f_2(\mathbf{x})$ be two probability densities on $E^n$. We say that $f_1(\mathbf{x})$ is more peaked than $f_2(\mathbf{x})$ if $f_1(\mathbf{x})d\mathbf{x}$ is more peaked than $f_2(\mathbf{x})d\mathbf{x}$.*

Intuitively speaking, if $f_1(\mathbf{x})$ is more peaked than $f_2(\mathbf{x})$, it means that $f_1(\mathbf{x})$ is larger than $f_2(\mathbf{x})$ in the vicinity of the origin. Let $\mu_1$ and $\mu_2$ be two probability measures on $E^{n_1}$ and $E^{n_2}$, respectively, and let $\mu_1 \otimes \mu_2$ denote the product of $\mu_1$ and $\mu_2$ on $E^{n_1+n_2}$. Clearly, $\mu_1 \otimes \mu_2$ is a probability measure on $E^{n_1+n_2}$.

**Lemma 1.2 (Kanter, 1977).** *Suppose that $\mu_{11}$, $\mu_{12}$, $\mu_{21}$, and $\mu_{22}$ are logconcave probability measures such that $\mu_{11}$ is more peaked than $\mu_{21}$ on $E^{n_1}$ and $\mu_{12}$ is more peaked than $\mu_{22}$ on $E^{n_2}$. Then $\mu_{11} \otimes \mu_{12}$ is more peaked than $\mu_{21} \otimes \mu_{22}$ on $E^{n_1+n_2}$.*

**Proof (A sketch).** First, by Lemma 1.1 it is easy to see that both $\mu_{11} \otimes \mu_{12}$ and $\mu_{21} \otimes \mu_{22}$ are logconcave.

Let $L_n$ denote the family of continuous nonnegative logconcave functions $f(\mathbf{x})$ on $E^n$ satisfying $f(-\mathbf{x}) = f(\mathbf{x})$ and

$$\lim_{\|\mathbf{x}\| \to \infty} f(\mathbf{x}) = 0.$$

By approximation, it can be shown (see Kanter, 1977 for the details[2]) that *a probability measure $\mu$ is more peaked than another probability measure $\nu$ if and only if*

$$\int_{E^n} f(\mathbf{x})\, d\mu \geq \int_{E^n} f(\mathbf{x})\, d\nu \tag{1.7}$$

*holds for all $f(\mathbf{x}) \in L_n$.*

Let $g(\mathbf{x}, \mathbf{y})$ be a logconcave function defined on $E^{n_1+n_2}$, where $\mathbf{x} \in E^{n_1}$ and $\mathbf{y} \in E^{n_2}$, and define

$$G(\mathbf{y}) = \int_{E^{n_1}} g(\mathbf{x}, \mathbf{y})\, d\mathbf{x}.$$

We proceed to show that $G(\mathbf{y})$ is logconcave on $E^{n_2}$. To this end, for $\mathbf{y}_1, \mathbf{y}_2 \in E^{n_2}$ and $0 < \lambda < 1$, we define

$$g_1(\mathbf{x}) = g(\mathbf{x}, \mathbf{y}_1), \quad g_2(\mathbf{x}) = g(\mathbf{x}, \mathbf{y}_2),$$

and

$$g_3(\mathbf{x}) = g(\mathbf{x}, \lambda\mathbf{y}_1 + (1-\lambda)\mathbf{y}_2).$$

Since $g(\mathbf{x}, \mathbf{y})$ is logconcave, we have

$$g_3(\mathbf{x}) \geq \sup_{\lambda\mathbf{x}_1 + (1-\lambda)\mathbf{x}_2 = \mathbf{x}} g_1(\mathbf{x}_1)^\lambda g_2(\mathbf{x}_2)^{1-\lambda}. \tag{1.8}$$

Therefore, by (1.6) we get

$$\begin{aligned} G(\lambda\mathbf{y}_1 + (1-\lambda)\mathbf{y}_2) &\geq \left(\int_{E^{n_1}} g(\mathbf{x}_1, \mathbf{y}_1)\, d\mathbf{x}_1\right)^\lambda \left(\int_{E^{n_1}} g(\mathbf{x}_2, \mathbf{y}_2)\, d\mathbf{x}_2\right)^{1-\lambda} \\ &= G(\mathbf{y}_1)^\lambda G(\mathbf{y}_2)^{1-\lambda}, \end{aligned} \tag{1.9}$$

which means that $G(\mathbf{y})$ is logconcave.

Assume that $d\mu_{11} = p(\mathbf{x})d\mathbf{x}$, where $p(\mathbf{x})$ is a logconcave function defined on $E^{n_1}$. For any $f(\mathbf{x}, \mathbf{y}) \in L_{n_1+n_2}$, it follows by the previous statement and the assumption on $\mu_{11}$ that

$$F(\mathbf{y}) = \int_{E^{n_1}} f(\mathbf{x}, \mathbf{y})\, p(\mathbf{x})\, d\mathbf{x} \in L_{n_1+n_2}.$$

Then, by (1.7) we get

$$\begin{aligned} \int_{E^{n_1+n_2}} f(\mathbf{x}, \mathbf{y})\, d\mu_{11} d\mu_{12} &= \int_{E^{n_2}} F(\mathbf{y})\, d\mu_{12} \\ &\geq \int_{E^{n_2}} F(\mathbf{y})\, d\mu_{22} \\ &= \int_{E^{n_1+n_2}} f(\mathbf{x}, \mathbf{y})\, d\mu_{11} d\mu_{22}. \end{aligned}$$

[2] Kanter did not use the notation of peakedness, which was introduced by Birnbaum, (1948). Instead, he used its complementary relation, the *symmetric dominance*.

Thus, $\mu_{11} \otimes \mu_{12}$ is more peaked than $\mu_{11} \otimes \mu_{22}$. Similarly, we can get that $\mu_{11} \otimes \mu_{22}$ is more peaked than $\mu_{21} \otimes \mu_{22}$ and finally $\mu_{11} \otimes \mu_{12}$ is more peaked than $\mu_{21} \otimes \mu_{22}$. □

It is easy to see that $\chi(\overline{B^n}, \mathbf{x})$ is a logconcave probability density function on $E^n$. On the other hand, since

$$\begin{aligned}\int_{E^n} e^{-\pi\|\mathbf{x}\|^2} d\mathbf{x} &= n\omega_n \int_0^\infty e^{-\pi r^2} r^{n-1} dr \\ &= \frac{n\omega_n}{2\pi^{\frac{n}{2}}} \int_0^\infty e^{-x} x^{\frac{n}{2}-1} dx \\ &= \frac{\frac{n}{2} \cdot \Gamma(\frac{n}{2})}{\Gamma(\frac{n}{2}+1)} = 1, \end{aligned} \tag{1.10}$$

we can verify that $e^{-\pi\|\mathbf{x}\|^2}$ is a logconcave probability density function as well.

**Lemma 1.3 (Vaaler, 1979).** $\chi(\overline{B^n}, \mathbf{x})$ *is more peaked than* $e^{-\pi\|\mathbf{x}\|^2}$.

**Proof.** We express the points in polar coordinates $\mathbf{x} = r\mathbf{x}'$, where $r = \|\mathbf{x}\|$ and $\mathbf{x}' \in \partial(B^n)$. Then, for any centrally symmetric convex body $C$, we have

$$\int_C e^{-\pi\|\mathbf{x}\|^2} d\mathbf{x} = \int_{\partial(B^n)} \int_0^\infty \chi(C, r\mathbf{x}') e^{-\pi r^2} r^{n-1} dr d\mathbf{x}'. \tag{1.11}$$

Let $\mathbf{x}'$ be a fixed point on $\partial(B^n)$. Since both $C$ and $\overline{B^n}$ are convex, we have either

$$\chi(C, r\mathbf{x}') \leq \chi(\overline{B^n}, r\mathbf{x}') \tag{1.12}$$

or

$$\chi(C, r\mathbf{x}') \geq \chi(\overline{B^n}, r\mathbf{x}') \tag{1.13}$$

for all $0 \leq r < \infty$. If (1.12) holds at $\mathbf{x}'$, then we have

$$\begin{aligned}\int_0^\infty \chi(C, r\mathbf{x}')\, e^{-\pi r^2} r^{n-1} dr &\leq \int_0^\infty \chi(C, r\mathbf{x}')\, r^{n-1} dr \\ &= \int_0^\infty \chi(C, r\mathbf{x}')\, \chi(\overline{B^n}, r\mathbf{x}')\, r^{n-1} dr. \end{aligned} \tag{1.14}$$

If (1.13) holds at $\mathbf{x}'$, then we have

$$\begin{aligned}\int_0^\infty \chi(C, r\mathbf{x}')\, e^{-\pi r^2} r^{n-1} dr &\leq \int_0^\infty e^{-\pi r^2} r^{n-1} dr \\ &= \frac{\pi^{-\frac{n}{2}} \Gamma(\frac{n}{2}+1)}{n} \\ &= \int_0^\infty \chi(\overline{B^n}, r\mathbf{x}')\, r^{n-1} dr \\ &= \int_0^\infty \chi(C, r\mathbf{x}')\, \chi(\overline{B^n}, r\mathbf{x}')\, r^{n-1} dr. \end{aligned} \tag{1.15}$$

As a conclusion, by (1.11), (1.14), and (1.15) we get

$$\begin{aligned}\int_C e^{-\pi\|\mathbf{x}\|^2} d\mathbf{x} &\le \int_{\partial(B^n)} \int_0^\infty \chi(C, r\mathbf{x}')\, \chi(\overline{B^n}, r\mathbf{x}')\, r^{n-1} dr d\mathbf{x}' \\ &= \int_C \chi(\overline{B^n}, \mathbf{x})\, d\mathbf{x}.\end{aligned}$$

Thus, $\chi(\overline{B^n}, \mathbf{x})$ is more peaked than $e^{-\pi\|\mathbf{x}\|^2}$. Lemma 1.3 is proved. □

Now we can prove Vaaler's theorem.

**Proof of Theorem 1.1.** If $k = n$, then the theorem is trivial. So, we assume that $j = n - k$ is positive.

Let $E^k$ denote the $k$-dimensional subspace of $E^n$ spanned by the rows of $A$ and let $E^j$ denote the $j$-dimensional subspace of $E^n$ which is orthogonal to $E^k$. Then we may write a point of $E^n$ as $\mathbf{w} = (\mathbf{u}, \mathbf{v})$, where $\mathbf{u} \in E^k$ and $\mathbf{v} \in E^j$. Let $B$ be a $j \times n$ matrix such that its $j$ rows form an orthonormal basis in $E^j$. Then we define an $n \times n$ matrix

$$T = \begin{pmatrix} A \\ B \end{pmatrix}.$$

For $\epsilon > 0$, we define

$$H_\epsilon = \left\{(\mathbf{u}, \mathbf{v}) \in E^n : \max_{1 \le i \le j} |v_i| \le \tfrac{\epsilon}{2}\right\}$$

and

$$H'_\epsilon = \left\{\mathbf{v} \in E^j : \max_{1 \le i \le j} |v_i| \le \tfrac{\epsilon}{2}\right\}.$$

Clearly $H_\epsilon$ is a closed centrally symmetric subset of $E^n$.

By Lemma 1.2 and Lemma 1.3, we have

$$\int_{H_\epsilon} e^{-\pi\|\mathbf{w}T\|^2} d\mathbf{w} \le \int_{H_\epsilon} \chi(D, \mathbf{w}T)\, d\mathbf{w}. \tag{1.16}$$

Multiplying both sides of (1.16) by $\frac{1}{\epsilon^j}$ and writing $H_\epsilon = E^k \oplus H'_\epsilon$, we get

$$\frac{1}{\epsilon^j} \int_{E^k} \int_{H'_\epsilon} e^{-\pi\|\mathbf{u}A+\mathbf{v}B\|^2} d\mathbf{v} d\mathbf{u} \le \frac{1}{\epsilon^j} \int_{E^k} \int_{H'_\epsilon} \chi(D, \mathbf{u}A + \mathbf{v}B)\, d\mathbf{v} d\mathbf{u}. \tag{1.17}$$

By the orthogonality assumption, we get

$$\|\mathbf{u}A + \mathbf{v}B\|^2 = \|\mathbf{u}A\|^2 + \|\mathbf{v}B\|^2$$

and therefore

$$\begin{aligned}
&\lim_{\epsilon\to 0}\frac{1}{\epsilon^j}\int_{E^k}\int_{H'_\epsilon} e^{-\pi\|\mathbf{u}A+\mathbf{v}B\|^2}d\mathbf{v}d\mathbf{u}\\
&\quad=\lim_{\epsilon\to 0}\frac{1}{\epsilon^j}\int_{H'_\epsilon} e^{-\pi\|\mathbf{v}B\|^2}d\mathbf{v}\cdot\int_{E^k} e^{-\pi\|\mathbf{u}A\|^2}d\mathbf{u}\\
&\quad=\int_{E^k} e^{-\pi\|\mathbf{u}A\|^2}d\mathbf{u}\\
&\quad=k\omega_k\,|AA'|^{-\frac{1}{2}}\int_0^\infty e^{-\pi r^2}r^{k-1}dr\\
&\quad=|AA'|^{-\frac{1}{2}}, \qquad (1.18)
\end{aligned}$$

where the last step follows from (1.10). On the other hand, we have

$$\begin{aligned}
&\lim_{\epsilon\to 0}\frac{1}{\epsilon^j}\int_{E^k}\int_{H'_\epsilon}\chi(D,\mathbf{u}A+\mathbf{v}B)\,d\mathbf{v}d\mathbf{u}\\
&\quad=\int_{E^k}\left\{\lim_{\epsilon\to 0}\frac{1}{\epsilon^j}\int_{H'_\epsilon}\chi(D,\mathbf{u}A+\mathbf{v}B)\,d\mathbf{v}\right\}d\mathbf{u}\\
&\quad=_{E^k}\chi(D,\mathbf{u}A)\,d\mathbf{u}. \qquad (1.19)
\end{aligned}$$

By (1.17), (1.18), and (1.19), we finally get

$$|AA'|^{-\frac{1}{2}}\le\int_{E^k}\chi(D,\mathbf{u}A)\,d\mathbf{u}.$$

Theorem 1.1 is proved. □

## 1.3 Hensley's conjecture

Before Vaaler's complete proof for Good's conjecture, Hensley (1979) did prove the special case $k=n-1$; that is

$$v_{n-1}(I^n\bigcap H^{n-1})\ge 1$$

*holds for every $(n-1)$-dimensional subspace $H^{n-1}$ of $E^n$.* In the same paper, he also got an upper bound

$$v_{n-1}(I^n\bigcap H^{n-1})\le 5.$$

In addition, he made the following conjecture.

**Hensley's conjecture.** *For every $(n-1)$-dimensional subspace $H^{n-1}$ of $E^n$, we have*

$$v_{n-1}(I^n\bigcap H^{n-1})\le\sqrt{2}.$$

It is easy to see that

$$v_1(I^n \cap H^1) \le \sqrt{n}$$

holds for every straight line $H^1$. However, for $2 \le k \le n-2$, it turns out to be difficult even to make a conjecture for the best upper bounds for $v_k(I^n \cap H^k)$. Nevertheless, we can ask the following basic question.

What is the optimal upper bound for $v_k(I^n \cap H^k)$?

To answer this question, K. Ball proved the following theorems.

**Theorem 1.2 (Ball, 1989).** *For every k-dimensional hyperplane $H^k$ in $E^n$, we have*

$$v_k(I^n \cap H^k) \le \left(\tfrac{n}{k}\right)^{\frac{k}{2}}.$$

*Especially, if $k|n$, the upper bound is best possible.*

**Theorem 1.3 (Ball, 1989).** *For every k-dimensional hyperplane $H^k$ in $E^n$, we have*

$$v_k(I^n \cap H^k) \le 2^{\frac{n-k}{2}}.$$

*Especially, if $2(n-k) \le n$, the upper bound is optimal.*

**Remark 1.1.** Hensley's conjecture is proved by Theorem 1.3 as a special case $k = n-1$. However, when $k < \frac{n}{2}$ and $k$ is not a divisor of $n$, neither Theorem 1.2 nor Theorem 1.3 can answer the basic question.

**Remark 1.2.** When $k$ is comparatively small, Theorem 1.2 is stronger than Theorem 1.3; when $k$ is comparatively large, Theorem 1.3 is better. In addition, a routine computation shows that

$$\left(\frac{n}{k}\right)^{\frac{k}{2}} = \left(1 + \frac{n-k}{k}\right)^{\frac{k}{2}} \le e^{\frac{n-k}{2}}.$$

To prove these theorems, like the lower bound case, we need some deep results from analysis. For a unit vector $\mathbf{u}$, let $\mathbf{u} \otimes \mathbf{u}$ denote the orthogonal projection from $E^n$ to the one-dimensional subspace $\{\lambda \mathbf{u} : \lambda \in R\}$ and let $I_n$ denote the identity operator on $E^n$. In other words, writing as matrices

$$\mathbf{u} \otimes \mathbf{u} = \begin{pmatrix} u_1u_1 & u_1u_2 & \cdots & u_1u_n \\ u_2u_1 & u_2u_2 & \cdots & u_2u_n \\ \vdots & \vdots & \ddots & \vdots \\ u_nu_1 & u_nu_2 & \cdots & u_nu_n \end{pmatrix}$$

and

$$I_n = \begin{pmatrix} 1 & 0 & \cdots & 0 \\ 0 & 1 & \cdots & 0 \\ \vdots & \vdots & \ddots & \vdots \\ 0 & 0 & \cdots & 1 \end{pmatrix}.$$

To verify the matrix expression for $\mathbf{u} \otimes \mathbf{u}$, we can write an arbitrary vector $\mathbf{v}$ as $\alpha\mathbf{u} + \mathbf{w}$, where $\mathbf{w}$ is orthogonal to $\mathbf{u}$, and deduce from the expression that $\mathbf{v} \cdot (\mathbf{u} \otimes \mathbf{u}) = \alpha\mathbf{u}$. Clearly, we have

$$\operatorname{tr}(\mathbf{u} \otimes \mathbf{u}) = \langle \mathbf{u}, \mathbf{u} \rangle = 1, \tag{1.20}$$

where $\operatorname{tr}(A)$ indicates the *trace* of $A$.

**Lemma 1.4 (Brascamp and Lieb, 1976).** *Let $\mathbf{u}_1, \mathbf{u}_2, \ldots, \mathbf{u}_m$ be $m$ unit vectors in $E^n$ and $c_1, c_2, \ldots, c_m$ be $m$ positive numbers satisfying*

$$\sum_{i=1}^{m} c_i \mathbf{u}_i \otimes \mathbf{u}_i = I_n, \tag{1.21}$$

*where $m \geq n$ and $0 < c_i < 1$. Then, for every set of nonnegative integrable functions $\{f_1(x), f_2(x), \ldots, f_m(x)\}$, we have*

$$\int_{E^n} \prod_{i=1}^{m} f_i(\langle \mathbf{u}_i, \mathbf{x} \rangle)^{c_i} \, d\mathbf{x} \leq \prod_{i=1}^{m} \left( \int_{-\infty}^{\infty} f_i(x) \, dx \right)^{c_i}, \tag{1.22}$$

*where equality holds if $f_i(x)$ are identical Gaussian densities.*

**Proof (A sketch).** Let $g(x)$ be an integrable function over $(-\infty, \infty)$ and define $g^*(x)$ to be a positive even function, decreasing in $[0, \infty)$, and satisfying

$$v_1(\{x : |g(x)| \geq \lambda\}) = v_1(\{x : g^*(x) \geq \lambda\})$$

for all $\lambda \geq 0$. It can be shown (see Brascamp, Lieb, and Luttinger, 1974 for the details) that

$$\int_{-\infty}^{\infty} |g_i(x)| \, dx = \int_{-\infty}^{\infty} g_i^*(x) \, dx \tag{1.23}$$

and

$$\int_{E^n} \prod_{i=1}^{m} g_i(\langle \mathbf{u}_i, \mathbf{x} \rangle) \, d\mathbf{x} \leq \int_{E^n} \prod_{i=1}^{m} g_i^*(\langle \mathbf{u}_i, \mathbf{x} \rangle) \, d\mathbf{x}, \tag{1.24}$$

where $g_i(x)$ are $m$ arbitrary integrable functions. Similarly, let $g(\mathbf{x})$ be an integrable function in $E^d$ and define $g^*(\mathbf{x}) = h(\|\mathbf{x}\|)$ (the *Schwarz symmetrization* of $g(\mathbf{x})$) to be a positive function, where $h(r)$ is decreasing in $[0, \infty)$, satisfying

$$v_d\big(\{\mathbf{x} : |g(\mathbf{x})| \geq \lambda\}\big) = v_d\big(\{\mathbf{x} : g^*(\mathbf{x}) \geq \lambda\}\big)$$

for all $\lambda \geq 0$. Then, as with the previous case, it can be shown that

$$\int_{E^d} |g_i(\mathbf{x})|\, d\mathbf{x} = \int_{E^d} g_i^*(\mathbf{x})\, d\mathbf{x}$$

and

$$\int_{E^{nd}} \prod_{i=1}^{m} g_i(\langle \mathbf{u}_i, \mathbf{x}_1\rangle, \ldots, \langle \mathbf{u}_i, \mathbf{x}_d\rangle)\, d\mathbf{x}_1 \cdots d\mathbf{x}_d$$
$$\leq \int_{E^{nd}} \prod_{i=1}^{m} g_i^*(\langle \mathbf{u}_i, \mathbf{x}_1\rangle, \ldots, \langle \mathbf{u}_i, \mathbf{x}_d\rangle)\, d\mathbf{x}_1 \cdots d\mathbf{x}_d. \tag{1.25}$$

Based on (1.23) and (1.24), to prove the lemma, we may assume that $f_i(x)^{c_i}$ are nonnegative even integrable functions decreasing in $[0, \infty)$. In addition, for convenience, we assume that

$$f_i(x)^{c_i} = \sum_{j=1}^{l} \alpha_{ij}\, \chi_j(x), \tag{1.26}$$

where $l$ is a positive integer, $\alpha_{ij}$ are nonnegative numbers, and $\chi_j(x)$ are characteristic functions of some intervals $[-h_j, h_j]$. The general case can be handled by approximation.

For convenience, we abbreviate the left-hand side of (1.22) to $J(\mathbf{f})$. Then, for any positive integer $d$, we have

$$\int_{E^{nd}} \prod_{i=1}^{m} \prod_{j=1}^{d} f_i(\langle \mathbf{u}_i, \mathbf{x}_j\rangle)^{c_i} d\mathbf{x}_1 \cdots d\mathbf{x}_d = J(\mathbf{f})^d.$$

Let $F_i(\mathbf{x})$ denote the Schwarz symmetrization of $\prod_{j=1}^{d} f_i(x_j)^{c_i}$. Then it follows by (1.25) that

$$J(\mathbf{f})^d \leq \int_{E^{nd}} \prod_{i=1}^{m} F_i(\langle \mathbf{u}_i, \mathbf{x}_1\rangle, \ldots, \langle \mathbf{u}_i, \mathbf{x}_d\rangle)\, d\mathbf{x}_1 \cdots d\mathbf{x}_d. \tag{1.27}$$

In addition, by (1.26) and some combinatorial argument, it can be deduced that

$$F_i(\mathbf{x}) = \sum_{j=1}^{(d+1)^l} \beta_{ij}\, \chi(r_j B^d, \mathbf{x}), \tag{1.28}$$

where $\beta_{ij}$ and $r_j$ are suitable nonnegative numbers.

For convenience, we write

$$\|f(\mathbf{x})\|_p = \left(\int f(\mathbf{x})^{\frac{1}{p}}\, d\mathbf{x}\right)^p.$$

By (1.28) and *Hölder's inequality*, we get

$$\|f_i(x)^{c_i}\|_{c_i}^d = \|F_i(\mathbf{x})\|_{c_i} \geq (d+1)^{-c_i' l} \sum_{j=1}^{(d+1)^l} \beta_{ij} \|\chi(r_j B^d, \mathbf{x})\|_{c_i}, \tag{1.29}$$

where $c_i' = 1 - c_i$. It follows by (1.20) and (1.21) that $\sum_{i=1}^m c_i = n$. Thus it follows from (1.27), (1.28), and (1.29) that

$$\left( \frac{J(\mathbf{f})}{\prod_{i=1}^m \|f_i(x)^{c_i}\|_{c_i}} \right)^d \leq \frac{(d+1)^{(m-n)l} \sum_{j_1,\ldots,j_m} \beta_{1j_1} \cdots \beta_{mj_m} \int_{E^{nd}} \prod_{i=1}^m \chi(r_{j_i} B^d, \mathbf{y}_i)\, d\mathbf{x}_1 \cdots d\mathbf{x}_d}{\sum_{j_1,\ldots,j_m} \beta_{1j_1} \cdots \beta_{mj_m} \prod_{i=1}^m \|\chi(r_{j_i} B^d, \mathbf{x})\|_{c_i}}, \tag{1.30}$$

where $\mathbf{y}_i = (\langle \mathbf{u}_i, \mathbf{x}_1 \rangle, \ldots, \langle \mathbf{u}_i, \mathbf{x}_d \rangle)$.

Defining

$$\varphi_i(r, \mathbf{x}) = e^{\frac{c_i d}{2}\left(1 - \frac{\|\mathbf{x}\|^2}{r^2}\right)},$$

it can be shown that[3]

$$\chi(rB^d, \mathbf{x}) \leq \varphi_i(r, \mathbf{x}),$$

$$\|\varphi_i(r, \mathbf{x})\|_{c_i} \leq (3\sqrt{d})^{c_i} \|\chi(rB^d, \mathbf{x})\|_{c_i}, \tag{1.31}$$

and therefore by (1.21)

$$\frac{\int_{E^{nd}} \prod_{i=1}^m \chi(r_{j_i} B^d, \mathbf{y}_i)\, d\mathbf{x}_1 \cdots d\mathbf{x}_d}{\prod_{i=1}^m \|\chi(r_{j_i} B^d, \mathbf{x})\|_{c_i}} \leq (3\sqrt{d})^n \frac{\int_{E^{nd}} \prod_{i=1}^m \varphi_i(r_{j_i}, \mathbf{y}_i)\, d\mathbf{x}_1 \cdots d\mathbf{x}_d}{\prod_{i=1}^m \|\varphi_i(r_{j_i}, \mathbf{x})\|_{c_i}}$$

$$= (3\sqrt{d})^n.$$

Thus by (1.30) we get

$$\frac{J(\mathbf{f})}{\prod_{i=1}^m \|f_i(x)^{c_i}\|_{c_i}} \leq \lim_{d\to\infty} \sqrt[d]{(d+1)^{(m-n)l}(3\sqrt{d})^n} = 1.$$

The lemma is proved. □

---

[3] To show (1.31), we can estimate $\left(\|\varphi_i(r, \mathbf{x})\|_{c_i} / \|\chi(rB^d, \mathbf{x})\|_{c_i}\right)^{1/c_i}$ and applying Stirling's formula.

**Proof of Theorem 1.2.** Let $\{\mathbf{e}_1, \mathbf{e}_2, \ldots, \mathbf{e}_n\}$ be an orthonormal basis of $E^n$ and let $\Gamma$ denote the orthogonal projection on to $H^k$. For convenience, we write

$$c_j = \|\Gamma(\mathbf{e}_j)\|^2, \qquad \mathbf{u}_j = \frac{1}{\sqrt{c_j}}\Gamma(\mathbf{e}_j),$$

and assume $c_j \neq 0$, without loss of generality. Clearly, $\mathbf{u}_1, \mathbf{u}_2, \ldots, \mathbf{u}_n$ are unit vectors in $H^k$ satisfying

$$\sum_{j=1}^{n} c_j \mathbf{u}_j \otimes \mathbf{u}_j = \sum_{j=1}^{n} \Gamma(\mathbf{e}_j) \otimes \Gamma(\mathbf{e}_j) = \Gamma.$$

Thus we have

$$\sum_{j=1}^{n} c_j \mathbf{u}_j \otimes \mathbf{u}_j = I_k, \tag{1.32}$$

where $I_k$ is the identity on $H^k$.

Let $f_j(x)$ denote the characteristic function of the interval $[-\frac{1}{2\sqrt{c_j}}, \frac{1}{2\sqrt{c_j}}]$. Then we have

$$\begin{aligned} I^n \cap H^k &= \left\{\mathbf{x} \in H^k : |\langle \mathbf{x}, \mathbf{e}_j\rangle| \leq \tfrac{1}{2},\ 1 \leq j \leq n\right\} \\ &= \left\{\mathbf{x} \in H^k : |\langle \mathbf{x}, \mathbf{u}_j\rangle| \leq \tfrac{1}{2\sqrt{c_j}},\ 1 \leq j \leq n\right\} \end{aligned}$$

and, by Lemma 1.4

$$\begin{aligned} v_k(I^n \cap H^k) &= \int_{H^k} \prod_{j=1}^{n} f_j(\langle \mathbf{u}_j, \mathbf{x}\rangle)^{c_j}\, d\mathbf{x} \\ &\leq \prod_{j=1}^{n} \left(\int_{-\infty}^{\infty} f_j(x)\, dx\right)^{c_j} \\ &= \left(\prod_{j=1}^{n} c_j^{\,c_j}\right)^{-\frac{1}{2}}. \end{aligned}$$

By (1.32), it is easy to see that

$$\sum_{j=1}^{n} c_j = k.$$

Then it can be shown by routine analysis that

$$\prod_{j=1}^{n} c_j^{\,c_j} \geq \prod_{j=1}^{n} \left(\frac{\sum c_i}{n}\right)^{\frac{\sum c_i}{n}} = \left(\frac{k}{n}\right)^{k}.$$

Therefore, we have

$$v_k(I^n \cap H^k) \leq \left(\frac{n}{k}\right)^{\frac{k}{2}}.$$

On the other hand, we can prove that, if $k$ is a divisor of $n$, there is a $k$-dimensional hyperplane $H^k$ such that $I^n \bigcap H^k$ is a cube of volume $(\frac{n}{k})^{\frac{k}{2}}$. Theorem 1.2 is proved. □

To prove Theorem 1.3, besides Lemma 1.4, we need another basic result.

**Lemma 1.5 (Ball, 1986).** *If $\lambda \geq 2$, then*

$$\frac{1}{\pi}\int_{-\infty}^{\infty}\left|\frac{\sin t}{t}\right|^{\lambda} dt \leq \sqrt{\frac{2}{\lambda}},$$

*with equality being attained if and only if $\lambda = 2$.*

**Proof.** We divide the proof into two cases.

**Case 1.** $\lambda \geq 4$. When $|t| \leq 6/\sqrt{5}$, we have

$$\frac{\sin t}{t} = \sum_{j=0}^{\infty} \frac{(-t^2)^j}{(2j+1)!} < 1 - \frac{t^2}{6} + \frac{t^4}{120},$$

$$e^{-\frac{t^2}{6}} = \sum_{j=0}^{\infty} \frac{(-t^2)^j}{j!\, 6^j} > 1 - \frac{t^2}{6} + \frac{t^4}{72} - \frac{t^6}{1296},$$

$$\frac{t^4}{72} - \frac{t^6}{1296} - \frac{t^4}{120} \geq 0,$$

and therefore

$$0 \leq \frac{\sin t}{t} \leq e^{-\frac{t^2}{6}}.$$

Writing $m = 6/\sqrt{5}$ and applying $\lambda \geq 4$, we obtain

$$\begin{aligned}
\frac{1}{\pi}\int_{-\infty}^{\infty}\left|\frac{\sin t}{t}\right|^{\lambda} dt &< \frac{1}{\pi}\int_{-\infty}^{\infty} e^{-\frac{\lambda t^2}{6}}\, dt + \frac{2}{\pi}\int_{m}^{\infty} t^{-\lambda} dt \\
&= \frac{\sqrt{6}}{\sqrt{\lambda\pi}} + \frac{2}{\pi(\lambda-1)m^{\lambda-1}} \\
&\leq \left(\sqrt{\frac{3}{2\pi}} + \frac{2\sqrt{2}}{3\pi m^3}\right)\sqrt{\frac{2}{\lambda}} \\
&< \sqrt{\frac{2}{\pi}}.
\end{aligned}$$

**Case 2.** $2 \leq \lambda < 4$. Writing $\gamma = \frac{\lambda}{2} - 1$, we have $0 \leq \gamma < 1$. In this case, we proceed to show that

$$\frac{1}{\pi}\int_{-\infty}^{\infty}\left(\frac{\sin^2 t}{t^2}\right)^{1+\gamma} dt \leq \frac{1}{\sqrt{1+\gamma}}, \tag{1.33}$$

with equality being attained if and only if $\gamma = 0$.

For convenience, we write

$$\alpha_j = \frac{1}{\pi}\int_{-\infty}^{\infty} e^{-\frac{t^2}{\pi}}\left(1-e^{-\frac{t^2}{\pi}}\right)^j dt$$

and

$$\beta_j = \frac{1}{\pi}\int_{-\infty}^{\infty} \frac{\sin^2 t}{t^2}\left(1-\frac{\sin^2 t}{t^2}\right)^j dt.$$

It can be easily checked that $\beta_0 = 1$. Now we have

$$\begin{aligned}\left(\frac{\sin^2 t}{t^2}\right)^{1+\gamma} &= \frac{\sin^2 t}{t^2}\left(1-\left(1-\frac{\sin^2 t}{t^2}\right)\right)^{\gamma} \\ &= \frac{\sin^2 t}{t^2}\left(1+\sum_{j=1}^{\infty}\frac{\prod_{i=0}^{j-1}(i-\gamma)}{j!}\left(1-\frac{\sin^2 t}{t^2}\right)^j\right).\end{aligned}$$

Hence, by the *monotone convergence theorem*, we get

$$\frac{1}{\pi}\int_{-\infty}^{\infty}\left(\frac{\sin^2 t}{t^2}\right)^{1+\gamma} dt = 1+\sum_{j=1}^{\infty}\frac{\prod_{i=0}^{j-1}(i-\gamma)}{j!}\beta_j. \tag{1.34}$$

On the other hand, as with (1.34), we may write the right-hand side of (1.33) as

$$\begin{aligned}\frac{1}{\sqrt{1+\gamma}} &= \frac{1}{\pi}\int_{-\infty}^{\infty} e^{-\frac{(1+\gamma)t^2}{\pi}} dt \\ &= 1+\sum_{j=1}^{\infty}\frac{\prod_{i=0}^{j-1}(i-\gamma)}{j!}\alpha_j.\end{aligned}$$

Since $\prod_{i=0}^{j-1}(i-\gamma) \le 0$, to prove the lemma it is sufficient to show that $\alpha_j < \beta_j$ for $j = 1, 2, \ldots$.

Computing the first six values of $\alpha_j$ and $\beta_j$ up to two decimal places, we get the following table.

| j | 1 | 2 | 3 | 4 | 5 | 6 |
|---|---|---|---|---|---|---|
| $\alpha_j$ | $0.29\cdots$ | $0.16\cdots$ | $0.11\cdots$ | $0.08\cdots$ | $0.07\cdots$ | $0.05\cdots$ |
| $\beta_j$ | $0.33\cdots$ | $0.22\cdots$ | $0.17\cdots$ | $0.15\cdots$ | $0.13\cdots$ | $0.12\cdots$ |

It is easy to check that, for $j \ge 7$

$$\frac{1}{\sqrt{2\pi j}} < \frac{1}{2\sqrt{\pi(j+1)}} - \frac{1}{2ej}.$$

Therefore to complete the proof it suffices to show that in this case

$$\alpha_j \le \frac{1}{\sqrt{2\pi j}} \tag{1.35}$$

and

$$\beta_j \ge \frac{1}{2\sqrt{\pi(j+1)}} - \frac{1}{2ej}. \tag{1.36}$$

First, it is routine to show that, for $0 < x \le 1$

$$|\log x| \ge 2(1-x)^2.$$

Hence we have

$$\begin{aligned}
\alpha_j &= \frac{1}{\pi}\int_{-\infty}^{\infty} e^{-\frac{t^2}{\pi}}\left(1-e^{-\frac{t^2}{\pi}}\right)^j dt \\
&= \frac{2}{\sqrt{\pi}}\int_0^{\infty} e^{-u^2}\left(1-e^{-u^2}\right)^j du \\
&= \frac{1}{\sqrt{\pi}}\int_0^1 \frac{(1-x)^j}{\sqrt{|\log x|}}\, dx \\
&\le \frac{1}{\sqrt{2\pi}}\int_0^1 (1-x)^{j-1} dx \\
&= \frac{1}{\sqrt{2\pi j}},
\end{aligned}$$

which is (1.35).

On the other hand, clearly we have

$$\beta_j \ge \frac{2}{\pi}\int_1^{\infty} \frac{\sin^2 t}{t^2}\left(1-\frac{1}{t^2}\right)^j dt.$$

It is easy to see that $\frac{1}{t^2}(1-\frac{1}{t^2})^j$ is increasing for $1 \le t \le \sqrt{j+1}$ and decreasing for $t \ge \sqrt{j+1}$. Hence, if $j \ge 6$, we have $\sqrt{j+1} > 1 + \frac{\pi}{2}$

$$\int_{1+\frac{\pi}{2}}^{\sqrt{j+1}} \frac{\sin^2 t}{t^2}\left(1-\frac{1}{t^2}\right)^j dt \ge \int_1^{\sqrt{j+1}-\frac{\pi}{2}} \frac{\cos^2 t}{t^2}\left(1-\frac{1}{t^2}\right)^j dt$$

and

$$\int_{\sqrt{j+1}}^{\infty} \frac{\sin^2 t}{t^2}\left(1-\frac{1}{t^2}\right)^j dt \ge \int_{\sqrt{j+1}+\frac{\pi}{2}}^{\infty} \frac{\cos^2 t}{t^2}\left(1-\frac{1}{t^2}\right)^j dt.$$

Therefore, we get

$$\begin{aligned}
&\int_1^\infty \frac{\sin^2 t}{t^2}\left(1-\frac{1}{t^2}\right)^j dt\\
&\geq \frac{1}{2}\int_1^\infty \frac{1}{t^2}\left(1-\frac{1}{t^2}\right)^j dt - \frac{1}{2}\int_{\sqrt{j+1}-\frac{\pi}{2}}^{\sqrt{j+1}+\frac{\pi}{2}} \frac{\cos^2 t}{t^2}\left(1-\frac{1}{t^2}\right)^j dt\\
&\geq \frac{1}{2}\int_1^\infty \frac{1}{t^2}\left(1-\frac{1}{t^2}\right)^j dt - \frac{1}{2}\max_t\left\{\frac{1}{t^2}\left(1-\frac{1}{t^2}\right)^j\right\}\cdot\int_0^\pi \cos^2 t\, dt\\
&= \frac{1}{2}\int_0^1 (1-u^2)^j du - \frac{\pi}{4(j+1)}\left(1-\frac{1}{j+1}\right)^j\\
&= \frac{4^j}{2(2j+1)}\binom{2j}{j}^{-1} - \frac{\pi}{4j}\left(1-\frac{1}{j+1}\right)^{j+1}\\
&\geq \frac{1}{4}\sqrt{\frac{\pi}{j+1}} - \frac{\pi}{4ej}
\end{aligned}$$

and finally

$$\beta_j \geq \frac{1}{2\sqrt{\pi(j+1)}} - \frac{1}{2ej},$$

which proves (1.36). The equality case is easy to check. The lemma is proved. □

**Proof of Theorem 1.3.** Assume the assertion inductively for sections of $I^{n-1}$. Let $\overline{H^k}$ denote the $(n-k)$-dimensional subspace of $E^n$ which is perpendicular with $H^k$. We consider two cases.

**Case 1.** *$\overline{H^k}$ has a unit vector* $\mathbf{u}$ *with* $u_1 > 1/\sqrt{2}$. Let $T$ denote the orthogonal projection on to $\overline{\mathbf{e}_1}$, let $T'$ denote the orthogonal projection on to $\overline{\mathbf{u}}$, and write

$$C = \left\{\mathbf{x} \in E^n : |x_j| \leq \tfrac{1}{2},\ 2 \leq j \leq n\right\}.$$

In this case $T(I^n \bigcap H^k)$ is a subset of a $k$-dimensional section of the $(n-1)$-dimensional unit cube $T(C)$ and therefore, by the inductive assumption, we get

$$v_k\big(T(I^n \bigcap H^k)\big) \leq 2^{\frac{n-k-1}{2}}. \tag{1.37}$$

Then $\overline{\mathbf{u}}$ has an orthogonal basis $\mathbf{a}_1, \mathbf{a}_2, \ldots, \mathbf{a}_{n-1}$, where $\mathbf{a}_1 = T'(\mathbf{e}_1)$ and $\mathbf{a}_j \in \overline{\mathbf{e}_1}$, $2 \leq j \leq n-1$. It is easy to see that

$$\|T(\mathbf{a}_j)\| = \begin{cases} \|u_1\mathbf{a}_1\| & \text{if } j = 1,\\ \|\mathbf{a}_j\| & \text{if } 2 \leq j \leq n-1.\end{cases}$$

Thus, since $H^k \subset \overline{\mathbf{u}}$, by (1.37), we get

$$v_k(I^n \cap H^k) \le \frac{1}{|u_1|} \cdot v_k(T(I^n \cap H^k)) \le 2^{\frac{n-k}{2}}.$$

**Case 2.** *$\overline{H^k}$ has no unit vector with a coordinate larger than* $1/\sqrt{2}$. Let $\Gamma$ denote the orthogonal projection on to $\overline{H^k}$, and write $\alpha_j = \|\Gamma(\mathbf{e}_j)\|$ and $\mathbf{u}_j = \frac{1}{\alpha_j}\Gamma(\mathbf{e}_j)$ for $1 \le j \le n$. Then

$$\alpha_j \le \frac{1}{\sqrt{2}}, \quad j = 1, 2, \ldots, n.$$

If some $\alpha_j$ is zero, then the problem reduces to that for $I^{n-1}$. Thus, we assume that $\alpha_j > 0$ for each $j$. For $\mathbf{v} \in \overline{H^k}$, we define

$$f(\mathbf{v}) = v_k(I^n \cap (H^k + \mathbf{v})).$$

In addition, let $x_1, x_2, \ldots, x_n$ be independent *random variables*, each uniformly distributed on $[-\frac{1}{2}, \frac{1}{2}]$, with respect to some probability. Then the random vector $\mathbf{x} = (x_1, x_2, \ldots, x_n) \in E^n$ induces Lebesgue measure on $I^n$ and $f(\mathbf{v})$ is the continuous probability density function of the random vector $\Gamma(\mathbf{x})$.

Let $E(\mathbf{x})$ denote the *expectation* of the random variable $\mathbf{x}$ and write

$$\phi(\mathbf{w}) = \int_{\overline{H^k}} e^{i\langle \mathbf{v}, \mathbf{w}\rangle} f(\mathbf{v})\, d\mathbf{v}.$$

Here the $i$ over $e$ is $\sqrt{-1}$. Then, by some basic results in probability theory, we have

$$\begin{aligned}
\phi(\mathbf{w}) &= E\left(e^{i\langle \mathbf{w}, \Gamma(\mathbf{x})\rangle}\right) = E\left(e^{i\sum_{j=1}^n x_j \langle \mathbf{w}, \Gamma(\mathbf{e}_j)\rangle}\right) \\
&= E\left(e^{i\sum_{j=1}^n x_j \alpha_j \langle \mathbf{w}, \mathbf{u}_j\rangle}\right) \\
&= \prod_{j=1}^n \frac{\sin \frac{1}{2}\alpha_j \langle \mathbf{w}, \mathbf{u}_j\rangle}{\frac{1}{2}\alpha_j \langle \mathbf{w}, \mathbf{u}_j\rangle}.
\end{aligned}$$

By the *standard Fourier inversion formula*

$$\begin{aligned}
v_k(I^n \cap H^k) = f(\mathbf{o}) &= \frac{1}{(2\pi)^{n-k}} \int_{\overline{H^k}} \phi(\mathbf{w})\, d\mathbf{w} \\
&= \frac{1}{(2\pi)^{n-k}} \int_{\overline{H^k}} \prod_{j=1}^n \frac{\sin \frac{1}{2}\alpha_j \langle \mathbf{w}, \mathbf{u}_j\rangle}{\frac{1}{2}\alpha_j \langle \mathbf{w}, \mathbf{u}_j\rangle}\, d\mathbf{w} \\
&= \frac{1}{\pi^{n-k}} \int_{\overline{H^k}} \prod_{j=1}^n \frac{\sin \alpha_j \langle \mathbf{w}, \mathbf{u}_j\rangle}{\alpha_j \langle \mathbf{w}, \mathbf{u}_j\rangle}\, d\mathbf{w} \\
&\le \frac{1}{\pi^{n-k}} \int_{\overline{H^k}} \prod_{j=1}^n \left| \frac{\sin \alpha_j \langle \mathbf{w}, \mathbf{u}_j\rangle}{\alpha_j \langle \mathbf{w}, \mathbf{u}_j\rangle} \right| d\mathbf{w} \\
&= \frac{1}{\pi^{n-k}} \int_{\overline{H^k}} \prod_{j=1}^n \phi_j(\langle \mathbf{w}, \mathbf{u}_j\rangle)\, d\mathbf{w},
\end{aligned} \tag{1.38}$$

where

$$\phi_j(t) = \left| \frac{\sin(\alpha_j t)}{\alpha_j t} \right|. \tag{1.39}$$

It is easy to check that

$$\sum_{j=1}^{n} \alpha_j^2 \mathbf{u}_j \otimes \mathbf{u}_j = I_{n-k}$$

and therefore

$$\sum_{j=1}^{n} \alpha_j^2 = n - k.$$

By Lemma 1.4 and (1.38), we get

$$v_k(I^n \cap H^k) \leq \frac{1}{\pi^{n-k}} \prod_{j=1}^{n} \left( \int_{-\infty}^{\infty} \phi_j(t)^{\frac{1}{\alpha_j^2}} dt \right)^{\alpha_j^2}. \tag{1.40}$$

Writing

$$p_j = \alpha_j^{-2} \geq 2$$

and applying Lemma 1.5 with (1.39) to (1.40), we get

$$\begin{aligned} v_k(I^n \cap H^k) &\leq \prod_{j=1}^{n} \left( \frac{1}{\pi} \int_{-\infty}^{\infty} \left| \frac{\sin(\alpha_j t)}{\alpha_j t} \right|^{p_j} dt \right)^{\frac{1}{p_j}} \\ &= \prod_{j=1}^{n} \left( \frac{1}{\alpha_j \pi} \int_{-\infty}^{\infty} \left| \frac{\sin x}{x} \right|^{p_j} dx \right)^{\frac{1}{p_j}} \\ &\leq \prod_{j=1}^{n} \left( \frac{1}{\alpha_j} \sqrt{\frac{2}{p_j}} \right)^{\frac{1}{p_j}} = \prod_{j=1}^{n} 2^{\frac{1}{2p_j}} \\ &= 2^{\frac{n-k}{2}}. \end{aligned}$$

The theorem is proved. □

For convenience, we write

$$\alpha(n,k) = \max \left\{ v_k(I^n \cap H^k) \right\},$$

where the maximum is over all $k$-dimensional hyperplanes. By Theorem 1.2 and Theorem 1.3, many values of $\alpha(n,i)$ are known when $n$ is relatively small. Let us end this section by listing them up to $n = 10$ as the following table.

| $k$ | 1 | 2 | 3 | 4 | 5 | 6 | 7 | 8 | 9 | 10 |
|---|---|---|---|---|---|---|---|---|---|---|
| $\alpha(3, k)$ | $\sqrt{3}$ | $\sqrt{2}$ | 1 | | | | | | | |
| $\alpha(4, k)$ | 2 | 2 | $\sqrt{2}$ | 1 | | | | | | |
| $\alpha(5, k)$ | $\sqrt{5}$ | ?? | 2 | $\sqrt{2}$ | 1 | | | | | |
| $\alpha(6, k)$ | $\sqrt{6}$ | 3 | $\sqrt{8}$ | 2 | $\sqrt{2}$ | 1 | | | | |
| $\alpha(7, k)$ | $\sqrt{7}$ | ?? | ?? | $\sqrt{8}$ | 2 | $\sqrt{2}$ | 1 | | | |
| $\alpha(8, k)$ | $\sqrt{8}$ | 4 | ?? | 4 | $\sqrt{8}$ | 2 | $\sqrt{2}$ | 1 | | |
| $\alpha(9, k)$ | 3 | ?? | $\sqrt{27}$ | ?? | 4 | $\sqrt{8}$ | 2 | $\sqrt{2}$ | 1 | |
| $\alpha(10, k)$ | $\sqrt{10}$ | 5 | ?? | ?? | $\sqrt{32}$ | 4 | $\sqrt{8}$ | 2 | $\sqrt{2}$ | 1 |

## 1.4 Additional remarks

About the cross sections of the $n$-dimensional unit cube there is another natural problem:

For a $k$-dimensional hyperplane $H^k$, what is the possible shape of $I^n \bigcap H^k$?

When $n = 3$ and $k = 2$, as was shown in Section 1.1, $I^3 \bigcap H^2$ is either a hexagon or a parallelogram. In higher dimensions, however, the situation is much more complicated.

Let $f_j(P)$ denote the number of the $j$-dimensional faces of a polytope $P$. By an argument similar to the case $n = 3$ and $k = 2$, it can be easily deduced that $f_{n-2}(I^n \bigcap H^{n-1})$ is either $2n - 2$ or $2n$. In addition, $f_{n-2}(I^n \bigcap H^{n-1}) = 2n - 2$ holds if and only if $I^n \bigcap H^{n-1}$ is a parallelopiped.

Let $\mathbf{v}$ be a vertex of $I^n$ and let $H$ denote the $(n-1)$-dimensional hyperplane which is perpendicular to $\mathbf{v}$ and passing $\mathbf{o}$. It is easy to see that half of the $2n$ facets of $I^n$ contain $\mathbf{v}$ and the other half contain $-\mathbf{v}$. In addition, there is no facet of $I^n$ containing both $\mathbf{v}$ and $-\mathbf{v}$. Let $F$ be a facet of $I^n$, we have

$$d(F) = \sqrt{n-1}, \tag{1.41}$$

where $d(X)$ denotes the diameter of $X$. On the other hand, if $\mathbf{v} \in F$ and $H \bigcap F = \emptyset$, then we have

$$d(F) \leq \sqrt{2} \cdot \frac{\sqrt{n}}{2} = \sqrt{\frac{n}{2}} < \sqrt{n-1},$$

which contradicts (1.41). Therefore $H$ intersects the relative interior of every facet of $I^n$ and hence

$$f_{n-2}(I^n \bigcap H) = 2n.$$

Especially, when $n = 4$, we can deduce that $I^4 \bigcap H$ is a three-dimensional cross polytope. However, we do not know what $I^n \bigcap H$ looks like when $n$ is large.

When $k \ll n$, the shapes of $I^n \bigcap H^k$ can be very different from each other. It can be a $k$-dimensional parallelopiped; on the other hand, it can be very spherical, as it was shown by Dvoretzky's well-known theorem (see Dvoretzky, 1961 or Zong, 1996). It was proved by Bárány and Lovász (1982) that *almost every $k$-dimensional cross section of $I^n$ has at least $2^k$ vertices.* However, we do not know any good bound for the number of $j$-dimensional faces of a $k$-dimensional cross section of $I^n$.

Let $E(n, k, j)$ denote the expected number of $j$-dimensional faces of a random $k$-dimensional cross section of $I^n$. Lonke (2000) recently proved the following results

$$E(n, k, 0) = 2^k \binom{n}{k} \sqrt{\frac{2k}{\pi}} \int_0^\infty e^{-\frac{kt^2}{2}} \mu_{n-k}(tI^{n-k})\, dt,$$

*where $\mu_{n-k}$ is the $(n-k)$-dimensional Gaussian probability measure, and*

$$E(n, n-k, n-j) \sim \frac{(2n)^{j-k}}{(j-k)!}$$

*for fixed $1 \le k < j$ as $n \to \infty$.* As a consequence of the first formula, for $n \to \infty$, we can deduce

$$E(n, k, 0) \sim \frac{2^k}{\sqrt{k}} (\pi \log n)^{\frac{(k-1)}{2}}$$

and especially

$$E(n, n-1, 0) \sim \frac{2^n \sqrt{n}}{\pi}.$$

# 2

# Projections

## 2.1 Introduction

Let $\mathbf{u}$ be a unit vector in $E^2$ and let $H$ be the one-dimensional subspace of $E^2$ which is perpendicular to $\mathbf{u}$. Let $I^2 \bigvee H$ denote the projection of $I^2$ on to $H$. It is easy to see that, for any $\mathbf{u}$, $I^2 \bigvee H$ is a segment. Without loss of generality, by symmetry, to search for bounds for $\ell(I^2 \bigvee H)$ we may assume that

$$\mathbf{u} = (\cos\theta, \sin\theta)$$

with $0 \leq \theta \leq \frac{\pi}{4}$. By elementary geometry (see Figure 2.1), we get

$$\ell(I^2 \bigvee H) = 2 \cdot \tfrac{\sqrt{2}}{2} \cdot \cos\left(\tfrac{\pi}{2} - \tfrac{\pi}{4} - \theta\right) = \sqrt{2} \cdot \sin\left(\tfrac{\pi}{4} + \theta\right)$$

and therefore

$$1 \leq \ell(I^2 \bigvee H) \leq \sqrt{2}, \tag{2.1}$$

where the lower bound can be attained if and only if $H$ is an axis of $E^2$ and the upper bound can be attained if and only if $H$ contains two antipodal vertices of $I^2$.

In three-dimensional space, as with the cross-section case, the situation is more complicated. We start with an easy case. Let $H^1$ denote a one-dimensional subspace of $E^3$, defined by a unit vector $\mathbf{u}$

$$H^1 = \{\lambda\mathbf{u} : \lambda \in R\},$$

and let $I^3 \bigvee H^1$ denote the orthogonal projection of $I^3$ on to $H^1$. Clearly, for any $H^1$, the projection $I^3 \bigvee H^1$ is a segment. Then it can be shown that

$$\begin{aligned}\ell(I^3 \bigvee H^1) &= 2 \max\left\{|\langle \mathbf{u}, \mathbf{x}\rangle| : \mathbf{x} \in I^3\right\} \\ &= 2 \max_{|x_i| \leq 1/2} \left\{|u_1x_1 + u_2x_2 + u_3x_3|\right\} \\ &= |u_1| + |u_2| + |u_3|.\end{aligned}$$

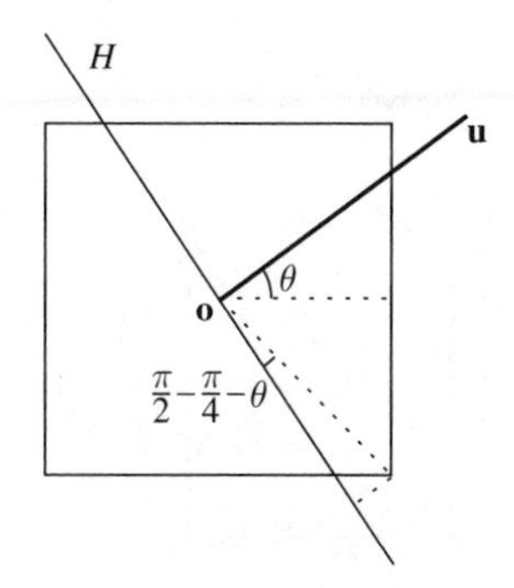


Figure 2.1

Therefore, we have

$$1 \leq \ell(I^3 \bigvee H^1) \leq \sqrt{3}, \tag{2.2}$$

where the lower bound can be attained if and only if $H^1$ is an axis of $E^3$ and the upper bound can be attained if and only if $H^1$ contains a pair of antipodal vertices of $I^3$.

Let $\mathbf{u}$ be a unit vector in $E^3$ and let $H^2$ denote the two-dimensional subspace of $E^3$ which is perpendicular to $\mathbf{u}$. In other words, $\mathbf{u}$ is a unit norm of $H^2$. For a subset $X$ of $E^3$, let $X \bigvee H^2$ denote its projection on to $H^2$. Clearly, $I^3 \bigvee H^2$ is a centrally symmetric polygon. For convenience, we write $\mathbf{w}_1 = (1, 0, 0)$, $\mathbf{w}_2 = (0, 1, 0)$, $\mathbf{w}_3 = (0, 0, 1)$, and define

$$F_1 = \left\{ \mathbf{x} \in I^3 : \langle \mathbf{x}, \mathbf{w}_1 \rangle = \tfrac{1}{2} \right\},$$

$$F_2 = \left\{ \mathbf{x} \in I^3 : \langle \mathbf{x}, \mathbf{w}_2 \rangle = \tfrac{1}{2} \right\},$$

and

$$F_3 = \left\{ \mathbf{x} \in I^3 : \langle \mathbf{x}, \mathbf{w}_3 \rangle = \tfrac{1}{2} \right\}.$$

By symmetry, to look for bounds for $v_2(I^3 \bigvee H^2)$ we may assume that $u_i \geq 0$ (see Figure 2.2). Then we have

$$I^3 \bigvee H^2 = (F_1 \bigvee H^2) \bigcup (F_2 \bigvee H^2) \bigcup (F_3 \bigvee H^2)$$

and therefore

$$\begin{aligned} v_2(I^3 \bigvee H^2) &= v_2(F_1 \bigvee H^2) + v_2(F_2 \bigvee H^2) + v_2(F_3 \bigvee H^2) \\ &= \langle \mathbf{u}, \mathbf{w}_1 \rangle \cdot v_2(F_1) + \langle \mathbf{u}, \mathbf{w}_2 \rangle \cdot v_2(F_2) + \langle \mathbf{u}, \mathbf{w}_3 \rangle \cdot v_2(F_3) \\ &= u_1 + u_2 + u_3. \end{aligned}$$

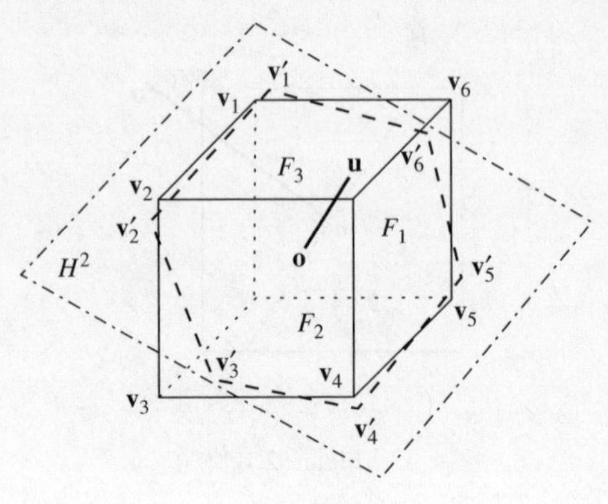


Figure 2.2

Since $u_1^2+u_2^2+u_3^2=1$, as with (2.2) we get

$$1 \leq v_2(I^3 \bigvee H^2) \leq \sqrt{3}, \tag{2.3}$$

where the lower bound can be attained if and only if $\mathbf{u}$ is an axis of $E^3$ and the upper bound can be attained if and only if $\mathbf{u}$ is in the direction of a vertex of $I^3$. In addition, it is easy to see that $I^3 \bigvee H^2$ is either a parallelogram or a hexagon.

Comparing with the cross sections of $I^3$, which were discussed in Section 1.1, it is easy to see that the cross sections and the projections are quite different. In fact, as we will see in higher dimensions, not only the results but also the proof methods are very different. While the key methods to deal with cross sections are analytic, the main ideas for projections are algebraic. In this chapter we will concentrate on the projections of $I^n$.

## 2.2 Lower bounds and upper bounds

Let $H^k$ denote a $k$-dimensional subspace of $E^n$ and let $I^n \bigvee H^k$ denote the orthogonal projection of $I^n$ on to $H^k$. Clearly, $I^n \bigvee H^k$ is a $k$-dimensional centrally symmetric polytope. As with the cross-section case, it is natural to ask the following question:

What are the best lower bound and the best upper bound for $v_k(I^n \bigvee H^k)$?

In 1986, G.D. Chakerian and P. Filliman did study this problem. For a lower bound, they proved the following result:

**Theorem 2.1 (Chakerian and Filliman, 1986).** *Whenever $1 \le k \le n-1$, for any $k$-dimensional orthogonal projection $I^n \bigvee H^k$ of $I^n$, we have*

$$v_k(I^n \bigvee H^k) \ge 1,$$

*where equality is attained if and only if $H^k$ is spanned by $k$ axes of $E^n$.*

**Proof.** It is easy to see that

$$I^n \bigcap H^k \subseteq I^n \bigvee H^k$$

and therefore

$$v_k(I^n \bigvee H^k) \ge v_k(I^n \bigcap H^k).$$

Then Theorem 2.1 follows from Corollary 1.1. □

**Remark 2.1.** It seems that Chakerian and Filliman were not aware of Vaaler's work on the lower bound of $v_k(I^n \bigcap H^k)$. Their original proof did use a different method which will be mentioned in the next section.

It is obvious that

$$I^n \subseteq \tfrac{\sqrt{n}}{2} B^n$$

and therefore

$$I^n \bigvee H^k \subseteq \tfrac{\sqrt{n}}{2} B^k$$

holds for every $k$-dimensional projection $I^n \bigvee H^k$ of $I^n$. Thus, we have

$$v_k(I^n \bigvee H^k) \le \omega_k \left(\tfrac{\sqrt{n}}{2}\right)^k. \tag{2.4}$$

Clearly this upper bound is not optimal, especially when $k$ is comparatively large. To improve this upper bound, we will prove the following results in this section.

**Theorem 2.2 (Chakerian and Filliman, 1986).**

$$v_k(I^n \bigvee H^k) \le \frac{\omega_{k-1}^k}{\omega_k^{k-1}} \left(\frac{n}{k}\right)^{k/2}. \tag{2.5}$$

**Theorem 2.3 (Chakerian and Filliman, 1986).**

$$v_k(I^n \bigvee H^k) \le \sqrt{\frac{n!}{(n-k)!\,k!}}. \tag{2.6}$$

**Remark 2.2.** For any fixed $k$ and relatively small $n$, among these three upper bounds (2.6) is the most efficient and (2.4) is the least efficient. On the other hand, for sufficiently large $n$, (2.5) is the most efficient and (2.4) is the least efficient. Nevertheless, for any fixed $k$ and sufficiently large $n$, the three upper bounds given by (2.4), (2.5), and (2.6) are essentially the same, all of the order $n^{k/2}$.

**Remark 2.3.** Let $c$ and $d$ be integers defined by $c = [n/(k+1)]$ and $d = n - c(k+1)$. For $i = 1, 2, \ldots, k+1$, let $\mathbf{p}_i = (p_{i1}, p_{i2}, \ldots, p_{in})$ be the vertex of $\overline{I^n}$ with coordinates

$$p_{ij} = \begin{cases} 1 & \text{if } (i-1)c+1 \le j \le ic, \\ 0 & \text{otherwise.} \end{cases}$$

Then $\mathbf{p}_1, \mathbf{p}_2, \ldots, \mathbf{p}_{k+1}$ are the vertices of a $k$-dimensional regular simplex $S$ of edge length $\sqrt{2c}$. It is easy to see that $S \subseteq \overline{I^n}$ and therefore there is a corresponding $I^n \bigvee H^k$ which contains a translate of $S$. Thus we have

$$v_k(I^n \bigvee H^k) \ge v_k(S) \sim \alpha_k (2c)^{k/2} \sim \alpha'_k n^{k/2},$$

where both $\alpha_k$ and $\alpha'_k$ are suitable constants depending only on $k$. It follows by this example that, though none of the three upper bounds given by (2.4), (2.5), and (2.6) is optimal in general, they do have the correct asymptotic order.

To prove these theorems, we need some basic results in convex geometry. Let $K$ and $K'$ be convex sets in $E^n$, it is known as *Steiner's formula* that

$$v_n(K + tK') = \sum_{i=0}^{n} \binom{n}{i} V_i(K, K') \cdot t^i \tag{2.7}$$

holds for all $t \ge 0$, where $V_i(K, K')$ are constants only depending on $K$, $K'$, and $i$. Especially

$$V_0(K, K') = v_n(K)$$

and

$$V_n(K, K') = v_n(K').$$

When both $K$ and $K'$ are polytopes, the formula can be proved by a routine argument. The general case can be deduced by approximation.

When $K' = B^n$, the number

$$W_i(K) = V_i(K, B^n)$$

is known as the $i$th *quermassintegral* of $K$. In particular, it is known that $n \cdot W_1(K)$ is the surface area of $K$, $W_n(K) = \omega_n$, and

$$n \cdot W_{n-1}(K) = \int_{\partial(B^n)} h(K, \mathbf{u})\, d\mathbf{u}, \tag{2.8}$$

where

$$h(K, \mathbf{u}) = \max\left\{\langle \mathbf{u}, \mathbf{x}\rangle : \mathbf{x} \in K\right\}.$$

Since

$$h(K+K', \mathbf{u}) = h(K, \mathbf{u}) + h(K', \mathbf{u})$$

holds for every $\mathbf{u} \in \partial(B^n)$, by (2.8) we get the following lemma.

**Lemma 2.1.** *For every pair of convex sets $K$ and $K'$ in $E^n$, we have*

$$W_{n-1}(K+K') = W_{n-1}(K) + W_{n-1}(K').$$

For a convex set $K \subset E^n$ and a number $\lambda \in [0, 1]$, we define

$$K_\lambda = \lambda K + (1-\lambda)B^n$$

and

$$f(\lambda) = v_n(K_\lambda)^{\frac{1}{n}} - \lambda \cdot v_n(K)^{\frac{1}{n}} - (1-\lambda) \cdot v_n(B^n)^{\frac{1}{n}}.$$

It follows by (2.7) that

$$\begin{aligned} v_n(K_\lambda) &= \lambda^n \cdot v_n\left(K + \tfrac{1-\lambda}{\lambda} B^n\right) \\ &= \sum_{i=0}^{n} \binom{n}{i} W_i(K) \cdot (1-\lambda)^i \lambda^{n-i} \end{aligned}$$

and hence, by a routine computation

$$\begin{aligned} f'(0) &= W_n(K)^{-\frac{n-1}{n}} \left(W_{n-1}(K) - W_n(K)^{\frac{n-1}{n}} W_0(K)^{\frac{1}{n}}\right) \\ &= \omega_n^{-\frac{n-1}{n}} \left(W_{n-1}(K) - \omega_n^{\frac{n-1}{n}} v_n(K)^{\frac{1}{n}}\right). \end{aligned}$$

It follows by the Brunn–Minkowski inequality that $f(\lambda)$ is a concave function in $[0, 1]$. On the other hand, it is easy to see that

$$f(0) = f(1) = 0.$$

Therefore, we have $f'(0) \geq 0$ and hence

$$W_{n-1}(K) - \omega_n^{\frac{n-1}{n}} v_n(K)^{\frac{1}{n}} \geq 0.$$

In other words, we have proved the following inequality.

**Lemma 2.2 (Urysohn's inequality).** *For every $n$-dimensional convex set $K$, we have*

$$v_n(K) \leq \frac{W_{n-1}^n(K)}{\omega_n^{n-1}}.$$

**Proof of Theorem 2.2.** Let $\{\mathbf{u}_1, \mathbf{u}_2, \ldots, \mathbf{u}_n\}$ be an orthonormal set in $E^n$ and let $[\mathbf{o}, \mathbf{u}_i]$ denote the segment with endpoints $\mathbf{o}$ and $\mathbf{u}_i$. Then $\mathbf{u}_1, \mathbf{u}_2, \ldots, \mathbf{u}_n$ generate a unit cube $C$ given by

$$C = \sum_{i=1}^{n} [\mathbf{o}, \mathbf{u}_i] = \left\{ \sum_{i=1}^{n} \lambda_i \mathbf{u}_i : 0 \leq \lambda_i \leq 1 \right\}.$$

Let $H^k$ be the $k$-dimensional subspace of $E^n$ defined by

$$H^k = \left\{ \mathbf{x} \in E^n : x_j = 0, \ j = k+1, \ldots, n \right\}, \tag{2.9}$$

and let $\overline{\mathbf{u}_i}$ denote the orthogonal projection of $\mathbf{u}_i$ on to $H^k$. For convenience, we identify $H^k$ with $E^k$ by disregarding the last $n-k$ coordinates. So, if $\mathbf{u}_i = (u_{i1}, u_{i2}, \ldots, u_{in})$, then $\overline{\mathbf{u}_i} = (u_{i1}, u_{i2}, \ldots, u_{ik}, 0, 0, \ldots, 0)$. Thus the projection of $C$ on to $H^k$ is a *zonotope* in $E^k$

$$C \bigvee H^k = \sum_{i=1}^{n} [\mathbf{o}, \overline{\mathbf{u}_i}] = \left\{ \sum_{i=1}^{n} \lambda_i \overline{\mathbf{u}_i} : 0 \leq \lambda_i \leq 1 \right\}. \tag{2.10}$$

Since the $n \times n$ matrix

$$U = \begin{pmatrix} u_{11} & u_{12} & \cdots & u_{1n} \\ u_{21} & u_{22} & \cdots & u_{2n} \\ \vdots & \vdots & \ddots & \vdots \\ u_{n1} & u_{n2} & \cdots & u_{nn} \end{pmatrix}$$

is orthogonal, letting $l_i$ denote the length of $[\mathbf{o}, \overline{\mathbf{u}_i}]$, we have

$$\begin{aligned} \sum_{i=1}^{n} l_i^2 &= \sum_{i=1}^{n} \|\overline{\mathbf{u}_i}\|^2 = \sum_{i=1}^{n} \left\{ \sum_{j=1}^{k} u_{ij}^2 \right\} \\ &= \sum_{j=1}^{k} \left\{ \sum_{i=1}^{n} u_{ij}^2 \right\} = \sum_{j=1}^{k} 1 = k. \end{aligned} \tag{2.11}$$

It is well known and trivial that

$$W_{k-1}([\mathbf{o}, \overline{\mathbf{u}_i}]) = \frac{\omega_{k-1}}{k} l_i.$$

Therefore, by Lemma 2.1, we have

$$W_{k-1}(C \bigvee H^k) = \sum_{i=1}^{n} W_{k-1}([\mathbf{o}, \overline{\mathbf{u}_i}]) = \frac{\omega_{k-1}}{k} \sum_{i=1}^{n} l_i. \tag{2.12}$$

Then, applying the Cauchy–Schwarz inequality to (2.12) and applying (2.11), we get

$$W_{k-1}(C\bigvee H^k) \le \frac{\omega_{k-1}}{k}\sqrt{n}\left(\sum_{i=1}^{n} l_i^2\right)^{1/2} = \omega_{k-1}\sqrt{\frac{n}{k}}.$$

Finally, by Lemma 2.2, we have

$$v_k(C\bigvee H^k) \le \frac{W_{k-1}^k(C\bigvee H^k)}{\omega_k^{k-1}} \le \frac{\omega_{k-1}^k}{\omega_k^{k-1}}\left(\frac{n}{k}\right)^{k/2}.$$

Theorem 2.2 is proved. □

Let $\{i_1, i_2, \ldots, i_k\}$ be a subset of $\{1, 2, \ldots, n\}$. Then the $k$ vectors $\overline{\mathbf{u}_{i_1}}$, $\overline{\mathbf{u}_{i_2}}, \ldots, \overline{\mathbf{u}_{i_k}}$ produce a parallelopiped

$$P_{i_1,i_2,\ldots,i_k} = \sum_{j=1}^{k}[\mathbf{o}, \overline{\mathbf{u}_{i_j}}] = \left\{\sum_{j=1}^{k}\lambda_j\overline{\mathbf{u}_{i_j}} : 0 \le \lambda_j \le 1\right\}$$

with $k$-dimensional volume

$$v_k(P_{i_1,i_2,\ldots,i_k}) = \left\| \begin{matrix} u_{i_1 1} & u_{i_1 2} & \cdots & u_{i_1 k} \\ u_{i_2 1} & u_{i_2 2} & \cdots & u_{i_2 k} \\ \vdots & \vdots & \ddots & \vdots \\ u_{i_k 1} & u_{i_k 2} & \cdots & u_{i_k k} \end{matrix} \right\|.$$

It is obvious that, for a given set $\{\overline{\mathbf{u}_1}, \overline{\mathbf{u}_2}, \ldots, \overline{\mathbf{u}_n}\}$, there are $\binom{n}{k}$ parallelopipeds of this kind. To prove Theorem 2.3, we need the following result about the volume of a zonotope. For convenience, we consider $C\bigvee H^k$ instead of $I^n\bigvee H^k$.

**Lemma 2.3 (Shephard, 1974).**

$$v_k(C\bigvee H^k) = \sum_{\{i_1,i_2,\ldots,i_k\}} v_k(P_{i_1,i_2,\ldots,i_k}),$$

*where the summation is over all possible subsets of* $\{1, 2, \ldots, n\}$ *of cardinality* $k$.

**Proof.** We apply induction on $n-k$. If $n-k=0$, there is nothing to prove. Assume that the statement is true for $n = m-1 \ge k$, we proceed to show it for $n = m$. Suppose that $\mathbf{u}_1, \mathbf{u}_2, \ldots, \mathbf{u}_{m-1}$, and $\mathbf{u}_m$ are $m$ orthonormal vectors, and write

$$C^i = \sum_{j=1}^{i}[\mathbf{o}, \mathbf{u}_j].$$

Then we have

$$C^m = C^{m-1} + [\mathbf{o}, \mathbf{u}_m]$$

and

$$C^m \bigvee H^k = C^{m-1} \bigvee H^k + [\mathbf{o}, \overline{\mathbf{u}_m}]. \tag{2.13}$$

If $\overline{\mathbf{u}_m} = \mathbf{o}$, then the lemma follows by (2.13) and the inductive assumption. If $\overline{\mathbf{u}_m} \neq \mathbf{o}$ and if $H^{k-1}$ is the subspace of $H^k$ which is perpendicular to $\overline{\mathbf{u}_m}$, then it follows by (2.13) and the inductive assumption that

$$\begin{aligned} v_k(C^m \bigvee H^k) &= v_k(C^{m-1} \bigvee H^k) + v_k(C^{m-1} \bigvee H^{k-1} + [\mathbf{o}, \overline{\mathbf{u}_m}]) \\ &= \sum_{\{i_1, i_2, \ldots, i_k\}^*} v_k(P_{i_1, i_2, \ldots, i_k}) + \sum_{\{i_1, i_2, \ldots, i_{k-1}\}^\star} v_k(P_{i_1, i_2, \ldots, i_{k-1}, m}) \\ &= \sum_{\{i_1, i_2, \ldots, i_k\}} v_k(P_{i_1, i_2, \ldots, i_k}), \end{aligned}$$

where $\{i_1, i_2, \ldots, i_k\}^*$ is over all possible subsets of $\{1, 2, \ldots, m-1\}$ of cardinality $k$, and $\{i_1, i_2, \ldots, i_{k-1}\}^\star$ is over all possible subsets of $\{1, 2, \ldots, m-1\}$ of cardinality $k-1$. The lemma is proved. □

**Proof of Theorem 2.3.** Let $C$, $\mathbf{u}_i$, $U$, and $P_{i_1, i_2, \ldots, i_k}$ be defined as above. Since $U$ is an orthogonal matrix, we have

$$U'U = UU' = I_n.$$

Thus, writing

$$A = \begin{pmatrix} u_{11} & u_{12} & \cdots & u_{1k} \\ u_{21} & u_{22} & \cdots & u_{2k} \\ \vdots & \vdots & \ddots & \vdots \\ u_{n1} & u_{n2} & \cdots & u_{nk} \end{pmatrix},$$

we can deduce

$$A'A = I_k.$$

By computing the determinants of both sides, it follows that

$$\sum_{\{i_1, i_2, \ldots, i_k\}} \left\| \begin{matrix} u_{i_1 1} & u_{i_1 2} & \cdots & u_{i_1 k} \\ u_{i_2 1} & u_{i_2 2} & \cdots & u_{i_2 k} \\ \vdots & \vdots & \ddots & \vdots \\ u_{i_k 1} & u_{i_k 2} & \cdots & u_{i_k k} \end{matrix} \right\|^2 = 1,$$

in other words

$$\sum_{\{i_1, i_2, \ldots, i_k\}} v_k(P_{i_1, i_2, \ldots, i_k})^2 = 1.$$

Then, by Lemma 2.3 and the Cauchy–Schwarz inequality, we get

$$\begin{aligned} v_k(C\bigvee H^k)^2 &= \left( \sum_{\{i_1,i_2,\ldots,i_k\}} v_k(P_{i_1,i_2,\ldots,i_k}) \right)^2 \\ &\leq \binom{n}{k} \left( \sum_{\{i_1,i_2,\ldots,i_k\}} v_k(P_{i_1,i_2,\ldots,i_k})^2 \right) \\ &= \binom{n}{k}. \end{aligned}$$

Theorem 2.3 is proved. □

## 2.3 A symmetric formula

The projections of a unit cube have the following symmetric property.

**Theorem 2.4 (McMullen, 1984 and Chakerian and Filliman, 1986).** *If $E^k$ and $E^{n-k}$ are orthogonal complementary subspaces of $E^n$, then*

$$v_k(I^n\bigvee E^k) = v_{n-k}(I^n\bigvee E^{n-k}).$$

**Remark 2.4.** The analogous statement of this theorem for cross sections is not always true. For example, defining

$$E^1 = \left\{\mathbf{x} \in E^3 :\ x_1 = x_2 = x_3\right\}$$

and

$$E^2 = \left\{\mathbf{x} \in E^3 :\ x_1 + x_2 + x_3 = 0\right\},$$

it follows by (1.3) that

$$v_1(I^3\bigcap E^1) \neq v_2(I^3\bigcap E^2).$$

Now let us introduce a basic result in linear algebra, from which we can easily deduce the above theorem.

**Lemma 2.4 (Jacobi's identity).** *Let $D$ be a subdeterminant of an $n \times n$ orthogonal matrix $U$ and let $D^*$ denote its algebraic complement, then*

$$|D^*| = |D|.$$

**Proof.** Assume that

$$D = U\begin{Bmatrix} i_1, i_2, \ldots, i_k \\ j_1, j_2, \ldots, j_k \end{Bmatrix},$$

where $i_l$ and $j_l$ denote respectively the rows and columns of the entries of $D$ in $U$, and write

$$D^* = U\begin{Bmatrix} i_1, i_2, \dots, i_k \\ j_1, j_2, \dots, j_k \end{Bmatrix}^*.$$

Let $V$ denote the $k \times n$ submatrix of the $n \times n$ unit matrix $I_n$ consisting of its $j_1, j_2, \dots, j_k$ rows and let $W$ denote the $n \times k$ submatrix of $I_n$ consisting of its $i_1, i_2, \dots, i_k$ columns.

Since $U$ is an orthogonal matrix, it is easy to see that

$$U' = U^{-1} = I_n U^{-1} I_n.$$

Therefore, we have

$$\begin{aligned} D = U\begin{Bmatrix} i_1, i_2, \dots, i_k \\ j_1, j_2, \dots, j_k \end{Bmatrix} &= U'\begin{Bmatrix} j_1, j_2, \dots, j_k \\ i_1, i_2, \dots, i_k \end{Bmatrix}. \\ &= \det(VU^{-1}W). \end{aligned} \tag{2.14}$$

Now we consider an $(n+k) \times (n+k)$ matrix

$$Q = \begin{pmatrix} U & W \\ V & 0_{k\times k} \end{pmatrix}.$$

By (2.14) we get

$$\begin{aligned} |\det(Q)| &= \left\| \begin{matrix} U & W \\ 0_{k\times n} & -VU^{-1}W \end{matrix} \right\| \\ &= \left|\det(VU^{-1}W)\right| \\ &= |D|. \end{aligned} \tag{2.15}$$

On the other hand, by *Laplace's expansion formula*, we get

$$\begin{aligned} |\det(Q)| &= \left| \det(I_k) \cdot Q\begin{Bmatrix} n+1, n+2, \dots, n+k \\ j_1, \quad j_2, \quad \dots, \quad j_k \end{Bmatrix}^* \right| \\ &= \left| U\begin{Bmatrix} i_1, i_2, \dots, i_k \\ j_1, j_2, \dots, j_k \end{Bmatrix}^* \right| = |D^*|. \end{aligned} \tag{2.16}$$

The lemma follows by (2.15) and (2.16). □

**Proof of Theorem 2.4.** As before, assume that $\{\mathbf{u}_1, \mathbf{u}_2, \dots, \mathbf{u}_n\}$ is an orthonormal set in $E^n$ and the considered unit cube is

$$C = \sum_{i=1}^{n} [\mathbf{o}, \mathbf{u}_i] = \left\{ \sum_{i=1}^{n} \lambda_i \mathbf{u}_i : 0 \le \lambda_i \le 1 \right\}.$$

By Lemmas 2.3 and 2.4, we have

$$\begin{aligned}
v_k(C\bigvee E^k) &= \sum_{\{i_1,i_2,\ldots,i_k\}} \left| U\begin{Bmatrix} i_1, i_2, \ldots, i_k \\ 1, 2, \ldots, k \end{Bmatrix} \right| \\
&= \sum_{\{i_1,i_2,\ldots,i_k\}} \left| U\begin{Bmatrix} i_1, i_2, \ldots, i_k \\ 1, 2, \ldots, k \end{Bmatrix}^* \right| \\
&= \sum_{\{j_1,j_2,\ldots,j_{n-k}\}} \left| U\begin{Bmatrix} j_1, & j_2, & \ldots, j_{n-k} \\ k+1, & k+2, & \ldots, \quad n \end{Bmatrix} \right| \\
&= v_{n-k}(C\bigvee E^{n-k}).
\end{aligned}$$

Theorem 2.4 is proved. □

**Remark 2.5.** Comparing Theorems 2.2 and 2.3 with Theorems 1.2 and 1.3, we can see that our knowledge about projections is not as good as that about cross sections.

Let $\beta(n, k)$ denote the maximum area of a $k$-dimensional projection of $I^n$. By Theorem 2.4 and Theorem 2.3, we have

$$\beta(n, 1) = \beta(n, n-1) = \sqrt{n}. \tag{2.17}$$

Besides this simple result, the only known values of $\beta(n, k)$ are

$$\beta(n, 2) = \beta(n, n-2) = \cot\left(\tfrac{\pi}{2n}\right). \tag{2.18}$$

Assume that

$$C = \sum_{j=1}^{n} [\mathbf{o}, \mathbf{u}_j]$$

and

$$P_2 = \sum_{j=1}^{n} [\mathbf{o}, \mathbf{u}'_j],$$

where $\mathbf{u}'_j$ is the projection of $\mathbf{u}_j$ on to a two-dimensional plane $E^2$. By (2.11) we can deduce that

$$\sum_{j=1}^{n} \|\mathbf{u}'_j\|^2 = 2.$$

Let $L$ denote the *perimeter* of $P_2$. Then we have

$$L^2 = \left(2\sum_{j=1}^{n} \|\mathbf{u}'_j\|\right)^2 \leq 4n\sum_{j=1}^{n} \|\mathbf{u}'_j\|^2 = 8n$$

and by the *isoperimetric inequality* for $2n$-gons in $E^2$ (see Fejes Tóth, 1964),

$$v_2(P_2) \leq \cot\left(\tfrac{\pi}{2n}\right),$$

where equality is attained if and only if $P_2$ is a regular $2n$-gon with edge length $\sqrt{2/n}$. On the other hand, we can construct such a projection. Let $u_1, u_2, \ldots, u_n$ be complex numbers defined by

$$u_j = u_{j1} + iu_{j2} = \sqrt{2/n} \cdot e^{(j-1)\pi i/n},$$

where $i = \sqrt{-1}$. For convenience, we may view them as points in $E^2$. Then we have

$$\begin{cases} \sum u_{j1}^2 + \sum u_{j2}^2 = \sum |u_j|^2 = 2 \\ \sum u_{j1}^2 - \sum u_{j2}^2 + 2i \sum u_{j1} u_{j2} = \sum u_j^2 = 0 \end{cases}$$

and therefore

$$\begin{cases} \sum u_{j1}^2 = \sum u_{j2}^2 = 1 \\ \sum u_{j1} u_{j2} = 0. \end{cases}$$

It is known in linear algebra that then we can construct an $n \times n$ unimodular matrix $U = (u_{jl})$. Writing $\mathbf{u}_j = (u_{j1}, u_{j2}, \ldots, u_{jn})$, it can be verified that $C = \sum \mathbf{u}_j$ is an $n$-dimensional unit cube and the projection of $C$ on to $E^2$ is exactly the regular $2n$-gon with vertices $\pm u_1, \pm u_2, \ldots, \pm u_n$. As a conclusion, (2.18) is proved.

Now, we end this section by listing the known values of $\beta(n, k)$ up to $n = 7$ as the following table.

| $k$ | 1 | 2 | 3 | 4 | 5 | 6 |
|---|---|---|---|---|---|---|
| $\beta(3, k)$ | $\sqrt{3}$ | $\sqrt{3}$ | | | | |
| $\beta(4, k)$ | 2 | $\cot\left(\frac{\pi}{8}\right)$ | 2 | | | |
| $\beta(5, k)$ | $\sqrt{5}$ | $\cot\left(\frac{\pi}{10}\right)$ | $\cot\left(\frac{\pi}{10}\right)$ | $\sqrt{5}$ | | |
| $\beta(6, k)$ | $\sqrt{6}$ | $\cot\left(\frac{\pi}{12}\right)$ | ?? | $\cot\left(\frac{\pi}{12}\right)$ | $\sqrt{6}$ | |
| $\beta(7, k)$ | $\sqrt{7}$ | $\cot\left(\frac{\pi}{14}\right)$ | ?? | ?? | $\cot\left(\frac{\pi}{14}\right)$ | $\sqrt{7}$ |

## 2.4 Combinatorial shapes

Defining

$$H^k = \left\{ \mathbf{x} \in E^n : \ x_i = 0 \ for \ i > k \right\},$$

it is easy to see that the projection of $I^n$ on to $H^k$ is a $k$-dimensional unit cube. On the other hand, according to Dvoretzky (1963) and Larman and Mani (1975), roughly speaking, $I^n$ has a $k$-dimensional projection almost spherical when $n$ is sufficiently large and $k$ is relatively small. In addition, the projection of $I^n$ does change continuously while the image space changes. Based on the above observations, we can ask the following question.

**Problem 2.2.** Determine the maximal (minimal) number of the $j$-dimensional faces of a $k$-dimensional projection of $I^n$.

So far no exact result to this problem is known. However, we do know its answer from the probability point of view.

Let $G(n, k)$ denote the *Grassmannian* of $k$-dimensional linear subspaces of $E^n$ with the usual topology. It is well known that there is a unique rotation invariant *probability measure* $\mu_k$ on $G(n, k)$. Let $P$ be an $n$-dimensional polytope and let $H^k \in G(n, k)$, then it is clear that $P\bigvee H^k$ is a $k$-dimensional polytope. For $j \in \{0, 1, \ldots, k-1\}$, let $F_j(P\bigvee H^k)$ denote the family of the $j$-dimensional faces of $P\bigvee H^k$ and let $f_j(P\bigvee H^k)$ denote the cardinality of $F_j(P\bigvee H^k)$. Clearly, $f_j(P\bigvee H^k)$ is a random variable of integer values and its expectation is

$$E(P, k, j) = \int_{G(n,k)} f_j(P\bigvee H^k)\, d\mu_k.$$

Let $F$ and $G$ be two faces of $P$ with $F \subseteq G$, let $\mathbf{v}$ be a relative interior point of $F$, and write

$$H = \{\lambda(\mathbf{x} - \mathbf{v}) + \mathbf{v} :\ \lambda \in R \text{ and } \mathbf{x} \in F\}$$

and

$$C = \{\lambda(\mathbf{y} - \mathbf{v}) + \mathbf{v} :\ \lambda \geq 0 \text{ and } \mathbf{y} \in G\}.$$

Clearly $H$ is a hyperplane and $C$ is a cone with $\mathbf{v}$ as its apex. Let $C'$ denote the set of outward normals to hyperplanes supporting $C$ at $H$ and assume that $G$ is $d$-dimensional, then we define

$$\beta(F, G) = \frac{v_d((B^d + \mathbf{v}) \cap C)}{v_d(B^d)}$$

and

$$\gamma(F, G) = \frac{v_d((B^d + \mathbf{v}) \cap C')}{v_d(B^d)}.$$

According to Grünbaum (1967) they are called the *internal angle* and the *external angle* of $G$ at $F$, respectively. By the definition, we have

$$\beta(F, F) = \gamma(F, F) = 1.$$

For convenience, if $F \not\subseteq G$, we write

$$\beta(F, G) = \gamma(F, G) = 0.$$

Based on the results in Grünbaum (1968), McMullen (1975), and Santaló (1952), Affentranger and Schneider (1992) deduced a general formula for $E(P, k, j)$. Namely

$$E(P, k, j) = 2 \sum_{s \geq 0} \sum_{F \in F_j(P)} \sum_{G \in F_{k-1-2s}(P)} \beta(F, G) \cdot \gamma(G, P). \tag{2.19}$$

It is observed by Böröczky and Henk (1999) that, for $F \in F_j(I^n)$ and $G \in F_l(I^n)$ with $F \subseteq G$, we have

$$\beta(F, G) = \beta(F, I^l) = 2^{j-l},$$

$$\gamma(F, I^n) = 2^{j-n},$$

$$\text{card}\{F_j(I^n)\} = 2^{n-j} \binom{n}{j},$$

and the number of the $l$-dimensional faces containing a fixed $j$-face is $\binom{n-j}{l-j}$. Therefore, applying (2.19) to $P = I^n$, we get the following result.

**Theorem 2.5.**

$$E(I^n, k, j) = 2 \binom{n}{j} \sum_{s \geq 0} \binom{n-j}{k-1-2s-j} \sim \frac{2n^{k-1}}{(k-1-j)!j!}.$$

# 3
# Inscribed simplices

## 3.1 Introduction

What is the maximum area of a triangle contained in the unit square $I^2$?

Let $T$ be such a triangle with vertices $\mathbf{v}_1$, $\mathbf{v}_2$, and $\mathbf{v}_3$. It is easy to see that $T$ is inscribed in $I^2$; in other words, all the three vertices are on the boundary of $I^2$. If, without loss of generality, $\mathbf{v}_1$ is a relative interior point of an edge of $I^2$, then, by moving $\mathbf{v}_1$ along the edge in a suitable direction until a vertex of $I^2$, we can get a new triangle with at least the same area. Consequently, $\mathbf{v}_1$, $\mathbf{v}_2$, and $\mathbf{v}_3$ can be three of the four vertices of $I^2$ and therefore

$$v_2(T) = \tfrac{1}{2}.$$

In fact, by a similar idea, we can deduce the following result about the $k$-dimensional maximal simplices contained in an $n$-dimensional unit cube.

**Theorem 3.1.** *For any integer $k$ with $1 \leq k \leq n$, there is a $k$-dimensional maximal simplex contained in $I^n$ that all its vertices are vertices of $I^n$.*

**Proof.** Let $S$ be a $k$-dimensional simplex in the $n$-dimensional Euclidean space with vertices $\mathbf{v}_1, \mathbf{v}_2, \ldots, \mathbf{v}_k$, and $\mathbf{v}_{k+1}$, let $[\mathbf{v}_1^*, \mathbf{v}_1^\star]$ be a segment containing $\mathbf{v}_1$ as a relative interior point, and let $S^*$ and $S^\star$ denote the corresponding simplices with $\mathbf{v}_1$ being replaced by $\mathbf{v}_1^*$ and $\mathbf{v}_1^\star$, respectively. First, let us show the following assertion.

**Assertion 3.1.** *Either $v_k(S^*)$ or $v_k(S^\star)$ is not smaller than $v_k(S)$.*

For convenience, we write

$$H = \left\{ \sum_{i=2}^{k+1} \lambda_i \mathbf{v}_i : \sum_{i=2}^{k+1} \lambda_i = 0 \right\}$$

and

$$H' = \left\{ \mathbf{u} \in E^n : \langle \mathbf{u}, \mathbf{v} \rangle = 0 \text{ for all } \mathbf{v} \in H \right\}.$$

In other words, $H$ is a $(k-1)$-dimensional subspace parallel with the hyperplane determined by $\mathbf{v}_2, \ldots, \mathbf{v}_k$, and $\mathbf{v}_{k+1}$, and $H'$ is the $(n-k+1)$-dimensional subspace which is perpendicular with $H$. Let $v(\mathbf{v}_2, \ldots, \mathbf{v}_{k+1})$ denote the $(k-1)$-dimensional measure of the simplex with vertices $\mathbf{v}_2, \ldots, \mathbf{v}_k$, and $\mathbf{v}_{k+1}$, and let $p(\mathbf{x})$ denote the projection of $\mathbf{x}$ on to $H'$. Then we have

$$v_k(S) = \tfrac{1}{k} \cdot \|p(\mathbf{v}_1 - \mathbf{v}_2)\| \cdot v(\mathbf{v}_2, \ldots, \mathbf{v}_{k+1}),$$

$$v_k(S^*) = \tfrac{1}{k} \cdot \|p(\mathbf{v}_1^* - \mathbf{v}_2)\| \cdot v(\mathbf{v}_2, \ldots, \mathbf{v}_{k+1}),$$

and

$$v_k(S^\star) = \tfrac{1}{k} \cdot \|p(\mathbf{v}_1^\star - \mathbf{v}_2)\| \cdot v(\mathbf{v}_2, \ldots, \mathbf{v}_{k+1}).$$

Clearly $p(\mathbf{v}_1 - \mathbf{v}_2)$ is contained in the segment $[p(\mathbf{v}_1^* - \mathbf{v}_2), p(\mathbf{v}_1^\star - \mathbf{v}_2)]$ and therefore one of the two ends is not a relative interior point of the $(n-k+1)$-dimensional ball of radius $\|p(\mathbf{v}_1 - \mathbf{v}_2)\|$ and center $\mathbf{o}$ in $H'$. Thus either $\|p(\mathbf{v}_1^* - \mathbf{v}_2)\|$ or $\|p(\mathbf{v}_1^\star - \mathbf{v}_2)\|$ is not smaller than $\|p(\mathbf{v}_1 - \mathbf{v}_2)\|$, which implies the assertion.

If $S$ is a $k$-dimensional simplex inscribed in $I^n$ and $\mathbf{v}_1$ is a relative interior point of a $d$-dimensional face of $I^n$, replacing $\mathbf{v}_1$ by some suitable boundary point of the face, then by Assertion 3.1 we can get a new inscribed simplex which is not smaller than $S$ in volume. Therefore the theorem can be proved by repeating this process for every vertex of the starting simplex for at most $n-1$ times. □

By this theorem, to search for the maximum volume of a $k$-dimensional simplex inscribed in a unit cube, it is sufficient to deal with the *vertex simplices*, the ones whose vertices are vertices of the unit cube as well, of the corresponding dimensions. Now, as an example, we can easily answer the following question.

**Question 3.1.** What is the maximum area (volume) of a triangle (tetrahedron) contained in a three-dimensional unit cube?

Let us start with the triangle case (see Figure 3.1). By Theorem 3.1, we consider a vertex triangle $T_2$. It is easy to see that an edge of $T_2$ can take only three possible lengths, 1, $\sqrt{2}$, and $\sqrt{3}$, and one of the three must be $\sqrt{2}$. Then, by comparing the distance from a vertex to such an edge, we can deduce that

$$v_2(T_2) \le \tfrac{\sqrt{3}}{2}, \tag{3.1}$$

where equality is attained if and only if $T_2$ is a regular one of edge length $\sqrt{2}$.

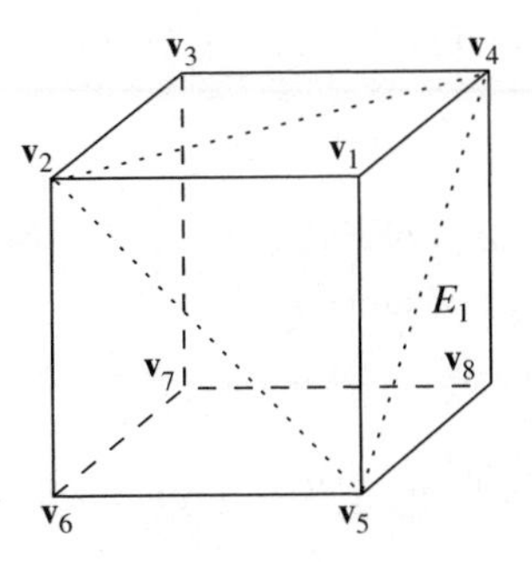


Figure 3.1

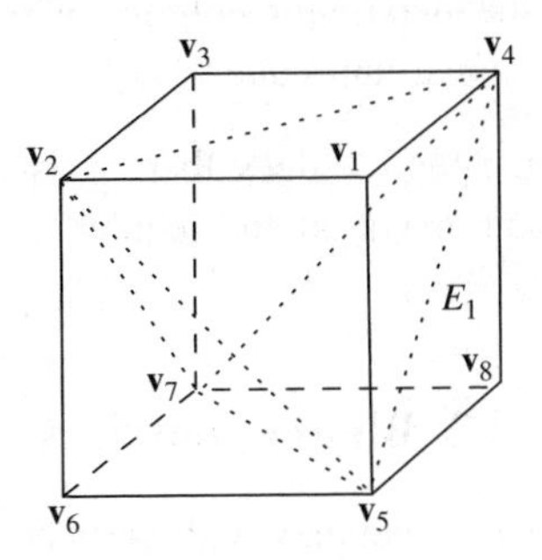


Figure 3.2

Let $T_3$ be a vertex tetrahedron of $I^3$. Since every vertex triangle contains at least one edge of length $\sqrt{2}$, the tetrahedron $T_3$ has two edges $E_1$ and $E_2$ of length $\sqrt{2}$. If $E_1$ and $E_2$ are co-planar (see Figure 3.2), by comparing the height of the possible tetrahedra with respect to the plane, we can deduce that

$$v_3(T_3) \leq 1 - 4 \cdot \tfrac{1}{3!} = \tfrac{1}{3}, \tag{3.2}$$

where equality holds if and only if $T_3$ is a regular one of edge length $\sqrt{2}$. If $T_3$ does not have two co-planar edges of length $\sqrt{2}$, then the two vertices of $T_3$, which are not the ends of $E_1$, must belong to $\{\mathbf{v}_1, \mathbf{v}_3, \mathbf{v}_6, \mathbf{v}_8\}$. In addition, one of them must be contained in the facet containing $E_1$ and the other must be contained in the opposite facet. Then we get

$$v_3(T_3) \leq \tfrac{1}{6}. \tag{3.3}$$

As a conclusion of (3.1), (3.2), and (3.3), we have proved the following result.

**Theorem 3.2.** *For every triangle $T_2$ contained in $I^3$, we have*

$$v_2(T_2) \le \tfrac{\sqrt{3}}{2},$$

*where equality holds if and only if $T_2$ is a regular triangle of edge length $\sqrt{2}$. For every tetrahedron $T_3$ contained in $I^3$, we have*

$$v_3(T_3) \le \tfrac{1}{3},$$

*where equality holds if and only if $T_3$ is a regular triangle of edge length $\sqrt{2}$.*

In this chapter we deal with problems of following types.

**Problem 3.1.** What is the maximum volume of a $k$-dimensional simplex contained in the $n$-dimensional unit cube $I^n$?

**Problem 3.2.** For which $n$ and $k$ does there exist a $k$-dimensional vertex simplex of $I^n$ which is both maximal and regular?

## 3.2 Binary matrices

Let $S^\star$ denote a $k$-dimensional simplex with vertices $\mathbf{v}_0 = (0, 0, \dots, 0)$, $\mathbf{v}_1 = (1, 0, \dots, 0), \dots, \mathbf{v}_k = (0, 0, \dots, 1)$ in the $k$-dimensional Euclidean space $E^k$. Then it is well known that

$$v_k(S^\star) = \int_{S^\star} d\mathbf{x} = \int \cdots \int_{\sum x_i \le 1,\ x_i \ge 0} dx_1 \cdots dx_k = \tfrac{1}{k!}.$$

Let $S$ be a $k$-dimensional simplex with vertices $\mathbf{a}_0 = (0, 0, \dots, 0)$, $\mathbf{a}_1 = (a_{11}, a_{12}, \dots, a_{1n}), \dots, \mathbf{a}_k = (a_{k1}, a_{k2}, \dots, a_{kn})$ in the $n$-dimensional Euclidean space $E^n$. For convenience, we assume that $\mathbf{a}_1, \mathbf{a}_2, \dots, \mathbf{a}_k$ are linearly independent; that is

$$v_k(S) > 0.$$

Let $H$ denote the $k$-dimensional subspace spanned by $\mathbf{a}_1, \mathbf{a}_2, \dots, \mathbf{a}_{k-1}$, and $\mathbf{a}_k$, and let $H'$ denote the $(n-k)$-dimensional subspace which is perpendicular with $H$. In other words

$$H = \left\{ \sum_{i=1}^{k} \lambda_i \mathbf{a}_i :\ \lambda_i \in R \right\}$$

and

$$H' = \left\{ \mathbf{x} \in E^n :\ \langle \mathbf{x}, \mathbf{y} \rangle = 0 \text{ for all } \mathbf{y} \in H \right\}. \tag{3.4}$$

Let $\{\mathbf{a}_{k+1}, \mathbf{a}_{k+2}, \ldots, \mathbf{a}_n\}$ be a basis of $H'$ such that

$$\langle \mathbf{a}_i, \mathbf{a}_j \rangle = \begin{cases} 1 & \text{if } i = j, \\ 0 & \text{otherwise} \end{cases} \tag{3.5}$$

holds for $k+1 \le i \le j \le n$. Then we define

$$S^* = \left\{ S + \sum_{i=k+1}^{n} \lambda_i \mathbf{a}_i : 0 \le \lambda_i \le 1 \right\}.$$

It is easy to see that

$$v_n(S^*) = v_k(S).$$

Let $A$ denote the $n \times n$ matrix with entries $a_{ij}$, let $A_k$ denote its $k \times n$ submatrix consisting of the first $k$ rows, and let $T$ denote the linear transformation determined by

$$T(\mathbf{a}_i) = \mathbf{e}_i, \quad i = 1, 2, \ldots, n,$$

where $\{\mathbf{e}_1, \mathbf{e}_2, \ldots, \mathbf{e}_n\}$ is an orthonormal basis of $E^n$. Thus we have

$$T(S^*) = \left\{ S^\star + \sum_{i=k+1}^{n} \lambda_i \mathbf{e}_i : 0 \le \lambda_i \le 1 \right\}$$

and therefore

$$\begin{aligned} v_k(S) = v_n(S^*) &= \int_{S^*} d\mathbf{x} = |\det(A)| \int_{T(S^*)} d\mathbf{x} \\ &= \tfrac{1}{k!} \cdot |\det(A)|. \end{aligned} \tag{3.6}$$

By (3.4) and (3.5) it follows that

$$AA' = \begin{pmatrix} A_k A_k' & 0 \\ 0 & I_{n-k} \end{pmatrix}$$

and therefore

$$\det(AA') = \det(A_k A_k'). \tag{3.7}$$

As a conclusion of (3.6) and (3.7) we have proved the following result.

**Theorem 3.3.**

$$v_k(S) = \tfrac{1}{k!} \sqrt{\det(A_k A_k')}.$$

Let $\mathbf{a}_0 = (0, 0, \ldots, 0)$, $\mathbf{a}_1 = (a_{11}, a_{12}, \ldots, a_{1n}), \ldots, \mathbf{a}_k = (a_{k1}, a_{k2}, \ldots, a_{kn})$ be $k+1$ vertices of $\overline{I^n}$, let $S$ denote the simplex determined by them, and let $A$ denote the $k \times n$ matrix with entries $a_{ij}$, where $i = 1, 2, \ldots, k$ and $j = 1, 2, \ldots, n$. Clearly, for any fixed $i$ and $j$, $a_{ij}$ is either zero or one, and therefore $A$ is a *binary matrix*. Then, by Theorem 3.3, Problem 3.1 can be restated as follows.

**Problem 3.1***. Determine the values of

$$\theta(n, k) = \max\left\{\det(AA')\right\},$$

where the maximum is over all $k \times n$ binary matrices.

For any simplex $S$ determined by $k+1$ vertices of $\overline{I^n}$, by Theorem 3.3, there is an integer $m$ between 0 and $\theta(n, k)$ satisfying

$$v_k(S) = \tfrac{1}{k!}\sqrt{m}. \tag{3.8}$$

Let $\psi(n, k, m)$ denote the number of the simplices satisfying (3.8) and let $\phi(n, k, m)$ denote the number of the simplices satisfying (3.8) and taking the origin as one of their vertices. It is easy to see that

$$\psi(n, k, m) = \frac{2^n}{k+1}\phi(n, k, m),$$

$$\sum_{i=0}^{\theta(n,k)} \psi(n, k, i) = \binom{2^n}{k+1},$$

and

$$\sum_{i=0}^{\theta(n,k)} \phi(n, k, i) = \binom{2^n - 1}{k}. \tag{3.9}$$

However, the following problem seems very hard.

**Problem 3.3.** Determine or estimate $\psi(n, k, m)$ and $\phi(n, k, m)$.

Let $S$ denote a simplex with $n+1$ vertices $\mathbf{a}_0 = (a_{01}, a_{02}, \ldots, a_{0n})$, $\mathbf{a}_1 = (a_{11}, a_{12}, \ldots, a_{1n}), \ldots, \mathbf{a}_n = (a_{n1}, a_{n2}, \ldots, a_{nn})$ in $E^n$. We define $a_{00} = a_{10} = \ldots = a_{n0} = 1$ and let $A$ denote the $(n+1) \times (n+1)$ matrix with entries $a_{ij}$, where $i, j = 0, 1, \ldots, n$. It is easy to see that

$$|\det(A)| = \left\| \begin{matrix} 1 & a_{01} & \cdots & a_{0n} \\ 0 & a_{11} - a_{01} & \cdots & a_{1n} - a_{0n} \\ \vdots & \vdots & \ddots & \vdots \\ 0 & a_{n1} - a_{01} & \cdots & a_{nn} - a_{0n} \end{matrix} \right\| = n! \cdot v_n(S). \tag{3.10}$$

On the other hand, when $S$ is a vertex simplex of $I^n$, multiplying $a_{ij}$ ($i = 0, 1, \ldots, n;\ j = 1, 2, \ldots, n$) by 2 we get an $(n+1) \times (n+1)$ matrix with $\pm 1$ entries. Thus, by Theorem 3.1, the special case $k = n$ of Problem 3.1 is equivalent to the following matrix problem.

**Problem 3.1***. What is the maximum determinant of an $(n+1)\times(n+1)$ matrix with $\pm 1$ entries?

**Remark 3.1.** When $m \neq 0$, it can be shown that the number of $k \times n$ binary matrices satisfying $\det(AA') = m$ is $k!\phi(n, k, m)$. Let $\tau(n, l)$ denote the number of the $n \times n$ binary matrices satisfying $\det(A) = l$ and let $\tau^*(n, l)$ denote the number of the $n \times n$ matrices with $a_{ij} = \pm 1$ and satisfying $\det(A) = l$. It is a challenging problem to estimate $\tau(n, l)$ and $\tau^*(n, l)$, especially $\tau(n, 0)$ and $\tau^*(n, 0)$. For related references we refer to Kahn, Komlós, and Szemerédi (1995).

**Remark 3.2 (Williamson, 1946).** Let $\theta_n$ denote the maximum determinant of an $n \times n$ binary matrix and let $\theta_n^*$ denote the maximum determinant of an $n \times n$ matrix of $\pm 1$ entries. By (3.10) it is easy to see that

$$\theta_{n+1}^* = 2^n \theta_n.$$

Therefore, to determine the $n$-dimensional maximal vertex simplex of $I^n$, we can deal with either $n \times n$ binary matrices or $(n+1)\times(n+1)$ matrices of $\pm 1$ entries.

An $n \times n$ matrix $A$ is called a *Hadamard matrix* if its entries are either 1 or $-1$ and satisfying

$$AA' = nI_n.$$

Hadamard matrices have many important applications, especially in applied mathematics (see Agaian, 1985 for detailed information).

Let $A$ be an $(n+1)\times(n+1)$ Hadamard matrix with entries $a_{ij}$, where both $i$ and $j$ run from 0 to $n$. If $a_{i0} = 1$ and

$$\mathbf{a}_i = (a_{i1}, a_{i2}, \ldots, a_{in})$$

for $i = 0, 1, \ldots, n$, then by definition it follows that

$$\langle \mathbf{a}_i, \mathbf{a}_j \rangle = \begin{cases} n & \text{if } i = j, \\ -1 & \text{otherwise,} \end{cases}$$

and thus

$$\|\mathbf{a}_i - \mathbf{a}_j\| = \sqrt{\langle \mathbf{a}_i - \mathbf{a}_j, \mathbf{a}_i - \mathbf{a}_j \rangle} = \sqrt{2n+2}$$

holds for every pair of distinct indices $i$ and $j$. Therefore, the simplex with vertices $\mathbf{a}_0, \mathbf{a}_1, \ldots, \mathbf{a}_{n-1}$, and $\mathbf{a}_n$ is regular. On the other hand, if $I^n$ contains an $n$-dimensional regular vertex simplex, then we can construct an $(n+1)\times(n+1)$ Hadamard matrix. As a conclusion, we have proved the following result.

**Theorem 3.4 (Grigorév, 1982).** *There is an n-dimensional regular vertex simplex in $I^n$ if and only if there is an $(n+1)\times(n+1)$ Hadamard matrix.*

## 3.3 Upper bounds

To approach Problem 3.1*, we will estimate the values of $\theta(n,k)$ in this section. Let $M_{k,n}$ denote the family of the $k\times n$ binary matrices and let $m{*}A$ denote the matrix obtained by concatenating a matrix $A$ in $m$ times. For convenience, let $J_k$ denote the $k\times k$ matrix with identical entries 1 and let $I_k$ denote the $k\times k$ unit matrix.

**Theorem 3.5 (Hudelson, Klee, and Larman, 1996 and Neubauer, Watkins, and Zeitlin, 1997).** *If $A\in M_{k,n}$, then*

$$\det(AA')\le\begin{cases}\frac{(k+1)^{k+1}n^k}{(4k)^k} & \text{if } k\equiv 1 \pmod 2,\\ \frac{(k+2)^k n^k}{4^k(k+1)^{k-1}} & \text{if } k\equiv 0 \pmod 2.\end{cases}$$

**Proof.** We will deduce this theorem by studying $((k+1)I_k-J_k)AA'$. It is easy to establish that

$$\det((k+1)I_k-J_k)=(k+1)^{k-1}$$

and therefore

$$\det(((k+1)I_k-J_k)AA')=(k+1)^{k-1}\det(AA'). \tag{3.11}$$

Let $\mathrm{tr}(B)$ denote the *trace* of an $n\times n$ matrix $B$; that is

$$\mathrm{tr}(B)=\sum_{i=1}^{n} b_{ii},$$

and let $\lambda_1(B),\lambda_2(B),\ldots,\lambda_n(B)$ denote the *eigenvalues* of $B$. It is known that

$$\det(B)=\prod_{i=1}^{n}\lambda_i(B) \tag{3.12}$$

and

$$\sum_{i=1}^{n}\lambda_i(B)=\mathrm{tr}(B). \tag{3.13}$$

If $B_1$ and $B_2$ are $n\times n$ matrices and $B_2$ is nonsingular, then the sets of the eigenvalues of $B_1B_2$ and $B_2B_1=B_2B_1B_2B_2^{-1}$ are identical. By a suitable limit process it can be shown that, if $B_1$ is an $n\times m$ matrix and $B_2$ is an $m\times n$ matrix, then the set of the nonzero eigenvalues of $B_1B_2$ is identical with that of $B_2B_1$.

By (3.12), (3.13), and the *arithmetic-geometric mean inequality*, we have

$$\det(((k+1)I_k - J_k)AA') \le \left(\tfrac{1}{k}\cdot \operatorname{tr}(((k+1)I_k - J_k)AA')\right)^k$$
$$= \left(\tfrac{1}{k}\cdot \operatorname{tr}(A'((k+1)I_k - J_k)A)\right)^k. \quad (3.14)$$

Suppose the $i$th column $\mathbf{c}$ of $A$ has $r$ ones and $k-r$ zeros. Then the $(i, i)$-entry of $A'((k+1)I_k - J_k)A$ is

$$\mathbf{c}'((k+1)I_k - J_k)\mathbf{c} = (k+1)\mathbf{c}'\mathbf{c} - \mathbf{c}'J_k\mathbf{c} = (k+1)r - r^2$$
$$= r(k+1-r).$$

If $A$ has $n_r$ columns each having exactly $r$ ones, then

$$\operatorname{tr}(A'((k+1)I_k - J_k)A) = \sum_{r=1}^{k} r(k+1-r)n_r. \quad (3.15)$$

Now we consider two cases.

**Case 1.** $k = 2l-1$. It is obvious that

$$r(k+1-r) \le l^2$$

and equality holds if and only if $r = l$. Thus we have

$$\sum_{r=1}^{k} r(k+1-r)n_r \le l^2 \sum_{r=1}^{k} n_r = l^2 n. \quad (3.16)$$

By (3.14), (3.15), and (3.16) we get

$$\det(((k+1)I_k - J_k)AA') \le \left(\frac{l^2 n}{k}\right)^k = \frac{(k+1)^{2k}n^k}{(4k)^k}$$

and, by (3.11)

$$\det(AA') \le \frac{(k+1)^{k+1}n^k}{(4k)^k}.$$

**Case 2.** $k = 2l$. It is obvious that

$$r(k+1-r) \le l(l+1)$$

and equality holds if and only if $r = l$ or $r = l+1$. Thus we have

$$\sum_{r=1}^{k} r(k+1-r)n_r \le l(l+1) \sum_{r=1}^{k} n_r = l(l+1)n. \quad (3.17)$$

By (3.14), (3.15), and (3.17) we get

$$\det(((k+1)I_k - J_k)AA') \le \left(\frac{l(l+1)n}{k}\right)^k = \frac{(k+2)^k n^k}{4^k}$$

and, by (3.11)

$$\det(AA') \le \frac{(k+2)^k n^k}{4^k (k+1)^{k-1}}.$$

The theorem is proved. □

By Theorem 3.3, one can restate this theorem in terms of inscribed simplex in $I^n$ as follows.

**Theorem 3.5*.** *For every $k$-dimensional simplex $S$ contained in $I^n$, we have*

$$v_k(S) \le \begin{cases} \frac{1}{k!2^k}\sqrt{\frac{(k+1)^{k+1}n^k}{k^k}} & \text{if } k \equiv 1 \pmod 2, \\ \frac{1}{k!2^k}\sqrt{\frac{(k+2)^k n^k}{(k+1)^{k-1}}} & \text{if } k \equiv 0 \pmod 2. \end{cases}$$

**Remark 3.3.** The first general upper bound in this setting was achieved by Hudelson, Klee, and Larman (1996). It is known (see Fejes Tóth, 1964) that the maximal $k$-dimensional simplices $S$ contained in an $n$-dimensional unit ball are regular. Based on this observation, they were able to prove

$$v_k(S) \le \frac{1}{k!2^k}\sqrt{\frac{(k+1)^{k+1}n^k}{k^k}},$$

where equality holds if $I^n$ contains a maximal $k$-dimensional regular vertex simplex. Let $G_j$ denote a $k \times \binom{k}{j}$ matrix such that its columns are $\binom{k}{j}$ distinct binary vectors and each has exactly $j$ ones. When $k = 2l-1$, for $n = m\cdot\binom{k}{l}$ and $A = m{*}G_l$, we indeed have

$$\theta(n,k) = \det(AA') = \frac{(k+1)^{k+1}n^k}{(4k)^k}. \tag{3.18}$$

When $k = 2l$, this upper bound was improved by Neubauer, Watkins, and Zeitlin (1997) into the listed form. In this case, by choosing $n = m\cdot\binom{2l+1}{l}$ and $A = m{*}[G_l, G_{l+1}]$, we get

$$\theta(n,k) = \det(AA') = \frac{(k+2)^k n^k}{4^k(k+1)^{k-1}}. \tag{3.19}$$

Clearly, as a general problem, Problem 3.1 as well as Problem 3.1* is still far from solved.

**Corollary 3.1 (Neubauer, Watkins, and Zeitlin, 1997).** *For every fixed $k$, we have*

$$\lim_{n\to\infty} \frac{\theta(n,k)}{n^k} = \begin{cases} \frac{(k+1)^{k+1}}{(4k)^k} & \text{if } k \equiv 1 \pmod 2, \\ \frac{(k+2)^k}{4^k(k+1)^{k-1}} & \text{if } k \equiv 0 \pmod 2. \end{cases}$$

**Proof.** Let $\gamma(n, k)$ denote the maximum volume of a $k$-dimensional simplex inscribed in the $n$-dimensional unit cube $I^n$. It is easy to see that

$$\gamma(n, k) < \gamma(n+1, k).$$

Therefore, by Theorem 3.3, we have

$$\theta(n, k) < \theta(n+1, k).$$

For convenience, we write

$$c = \begin{cases} \binom{k}{l} & \text{if } k = 2l-1, \\ \binom{k+1}{l} & \text{if } k = 2l, \end{cases}$$

$$\beta = \begin{cases} \frac{(k+1)^{k+1}}{(4k)^k} & \text{if } k = 2l-1, \\ \frac{(k+2)^k}{4^k(k+1)^{k-1}} & \text{if } k = 2l, \end{cases}$$

and

$$n = mc + r,$$

where $0 \leq r < c$. By (3.18) and (3.19) we have

$$\begin{aligned} \theta(mc, k) &= \beta\,(mc)^k \leq \theta(n, k) \leq \theta((m+1)c, k) \\ &= \beta\,((m+1)c)^k. \end{aligned}$$

Then it follows that

$$\frac{\theta(mc, k)}{n^k} \leq \frac{\theta(n, k)}{n^k} < \frac{\theta((m+1)c, k)}{n^k}$$

and therefore

$$\beta\left(\frac{mc}{mc+r}\right)^k \leq \frac{\theta(n, k)}{n^k} < \beta\left(\frac{(m+1)c}{mc+r}\right)^k.$$

The corollary follows. □

Now we discuss the most interesting and the most important case, $k = n$. Let us start with an upper bound discovered by Hadamard in 1893.

**Theorem 3.6 (Hadamard, 1893).** *For every $n \times n$ matrix $A$ with $\pm 1$ entries, we have*

$$\det(AA') \leq n^n,$$

*where equality holds if and only if $A$ is a Hadamard matrix.*

**Proof.** It is easy to see that

$$\mathrm{tr}(AA') = \sum_{i=1}^{n}\sum_{j=1}^{n} a_{ij}^2 = n^2.$$

Thus, by (3.12), (3.13), and the arithmetic-geometric mean inequality, we get

$$\det(AA') = \prod_{i=1}^{n} \lambda_i(AA') \leq \left(\tfrac{1}{n}\cdot \mathrm{tr}(AA')\right)^n = n^n,$$

where equality holds if and only if

$$\lambda_1(AA') = \lambda_2(AA') = \cdots = \lambda_n(AA') = n,$$

then $A$ is a Hadamard matrix. The theorem is proved. □

**Remark 3.4.** For every positive integer $k$, Sylvester (1867) proved that

$$S = \begin{pmatrix} 1 & 1 \\ 1 & -1 \end{pmatrix}$$

tensor with itself $k$ times gives a $2^k \times 2^k$ Hadamard matrix. Thus, the upper bound in Theorem 3.6 is tight for infinite values of $n$. However, it is not always tight. If there is an $n \times n$ Hadamard matrix, by Remark 3.2 we have

$$\theta_n^* = n^{n/2} = 2^{n-1}\theta_{n-1}.$$

Thus, since both $\theta_{n-1}$ and $\theta_n^*$ are integers, $n$ must be 1, 2 or a multiple of 4. This fact was discovered by Paley in 1933. On the other hand, up to now, we do not know if there is a $428 \times 428$ Hadamard matrix.

There are hundreds of papers dealing with Hadamard matrices. However, the following conjecture is still open.

**Conjecture 3.1 (Paley, 1933).** *There is an $n \times n$ Hadamard matrix whenever $n$ is a multiple of four.*

For convenience, we write

$$\theta^*(n, k) = \max\left\{\det(AA')\right\},$$

where the maximum is over all $k \times n$ matrices with $\pm 1$ entries. Based on Theorem 3.6 and Remark 3.4 it is reasonable to believe that

$$\lim_{n\to\infty} \frac{\theta^*(n, n)}{n^n} = 1.$$

However, as we will see from the following theorem, this is far from the truth.

**Theorem 3.7 (Barba, 1933; Ehlich, 1964a, 1964b; and Wojtas, 1964).**

$$\theta^*(n,n) \le \begin{cases} (2n-1)(n-1)^{n-1} & \text{if } n \equiv 1 \pmod 4, \\ 4(n-1)^2(n-2)^{n-2} & \text{if } n \equiv 2 \pmod 4, \\ \frac{4 \cdot 11^6}{7^7} n^7 (n-3)^{n-7} & \text{if } n \equiv 3 \pmod 4 \text{ and } n \ge 63. \end{cases}$$

The proof of this theorem, based on some basic results about symmetric matrices, is very complicated. We will introduce it case by case. Let us start with a couple of lemmas.

**Lemma 3.1.** *Let $C$ be an $n \times n$ symmetric matrix satisfying*

$$\begin{cases} c_{11} \ge c_{22} \ge \cdots \ge c_{nn} > 0, \\ |c_{in}| \ge c_{nn} \quad \text{for } i = 1, 2, \ldots, n, \end{cases}$$

*and the submatrix $C^* = (c_{ij})$, $i, j = 1, 2, \ldots, n-1$, is positive definite, then*

$$\det(C) \le c_{nn} \prod_{i=1}^{n-1} (c_{ii} - c_{nn}).$$

**Proof.** The assertion is obvious when $\det(C) \le 0$. If $\det(C) > 0$, then $C$ is positive definite. For convenience, we write

$$\mathbf{c} = (c_{n1}, c_{n2}, \ldots, c_{n,n-1})$$

and

$$D = C^* - \frac{1}{c_{nn}} \mathbf{c}' \mathbf{c}.$$

It is easy to see that

$$C \sim \begin{pmatrix} D & 0 \\ 0 & c_{nn} \end{pmatrix}.$$

Therefore $D$ is positive definite

$$\det(D) \le \prod_{i=1}^{n-1} d_{ii} = \prod_{i=1}^{n-1} \left( c_{ii} - \frac{c_{in}^2}{c_{nn}} \right)$$
$$\le \prod_{i=1}^{n-1} (c_{ii} - c_{nn})$$

and finally

$$\det(C) = c_{nn} \cdot \det(D) \le c_{nn} \prod_{i=1}^{n-1} (c_{ii} - c_{nn}).$$

The lemma is proved. □

**Lemma 3.2.** *Let $C$ be an $n \times n$ positive definite symmetric matrix satisfying*

$$\begin{cases} c_{11} \geq c_{22} \geq \cdots \geq c_{nn} \geq d, \\ |c_{ij}| \geq d \quad \textit{whenever } i \neq j. \end{cases}$$

*Then*

$$\det(C) \leq (c_{nn} + (n-1)d) \prod_{i=1}^{n-1} (c_{ii} - d),$$

*where equality can be attained.*

**Proof.** The assertion is obvious if $d = 0$. For convenience, we assume $d > 0$, write

$$C_k = \begin{pmatrix} c_{11} & c_{12} & \cdots & c_{1k} \\ c_{21} & c_{22} & \cdots & c_{2k} \\ \vdots & \vdots & \ddots & \vdots \\ c_{k1} & c_{k2} & \cdots & c_{kk} \end{pmatrix},$$

and let $D_k$ denote the $k \times k$ matrix obtained by replacing $c_{kk}$ by $d$ in $C_k$. Now we proceed to prove the lemma by induction.

When $n = 2$ the assertion is clear

$$\det(C_2) = c_{11}c_{22} - c_{12}^2 \leq (c_{11} - d)(c_{22} + d).$$

If the lemma is true for $n - 1$, then by Lemma 3.1 we get

$$\begin{aligned} \det(C_n) &= (c_{nn} - d) \cdot \det(C_{n-1}) + \det(D_n) \\ &\leq (c_{nn} - d)(c_{n-1,n-1} + (n-2)d) \prod_{i=1}^{n-2} (c_{ii} - d) + d \prod_{i=1}^{n-1} (c_{ii} - d) \\ &\leq (c_{n,n} + (n-2)d)(c_{n-1,n-1} - d) \prod_{i=1}^{n-2} (c_{ii} - d) + d \prod_{i=1}^{n-1} (c_{ii} - d) \\ &\leq (c_{nn} + (n-1)d) \prod_{i=1}^{n-1} (c_{ii} - d). \end{aligned}$$

The equality case can be easily verified and the lemma is proved. □

**Proof of the first case of Theorem 3.7.** Assume that $C = AA'$, where $A$ is an $n \times n$ matrix with $\pm 1$ entries. It is easy to see that $\det(C)$ keeps invariant if we multiply a whole row or a whole column of $A$ by $-1$. Therefore, since $n$ is odd, by suitable modification we can assume that the number $r_i$ of the $-1$ entries in $i$th row of $A$ is even. For convenience, we write

$$\beta_{ij} = \operatorname{card}\{l : a_{il} = a_{jl} = -1\}.$$

Then we have

$$c_{11} = c_{22} = \cdots = c_{nn} = n,$$

$$\begin{aligned} c_{ij} &= \sum_{l=1}^{n} a_{il}a_{jl} = n - (r_i + r_j - \beta_{ij}) + \beta_{ij} - (r_i + r_j - 2\beta_{ij}) \\ &= n - 2(r_i + r_j) + 4\beta_{ij} \equiv n \pmod 4, \end{aligned} \tag{3.20}$$

and therefore

$$|c_{ij}| \geq 1$$

holds for all indices $i$ and $j$. Applying Lemma 3.2 with $d = 1$, we get

$$\det(C) \leq (2n-1)(n-1)^{n-1}.$$

The first case of Theorem 3.7 is proved. □

To prove the second case of Theorem 3.7, we need the following technical lemma.

**Lemma 3.3 (Ehlich, 1964a).** *Let $\phi(C)$ denote the number of the elements $c_{ij} \equiv 0 \pmod 4$ in $C = AA'$, where $A$ is an $n \times n$ matrix with $\pm 1$ entries. If $n \equiv 2 \pmod 4$, then*

$$\phi(C) \leq n^2/2.$$

**Proof.** Let $r_i$ denote the number of the $-1$ entries in the $i$th row of $A$ and write

$$\beta_{ij} = \text{card}\{l : a_{il} = a_{jl} = -1\},$$

then we have

$$c_{ij} = \sum_{l=1}^{n} a_{il}a_{jl} = n - 2(r_i + r_j) + 4\beta_{ij}$$

and therefore

$$c_{ij} \equiv 0 \pmod 2$$

for all indices $i, j = 1, \ldots, n$. In addition, since there is no integer triple $\{r_i, r_j, r_l\}$ satisfying

$$r_i + r_j \equiv r_j + r_l \equiv r_i + r_l \equiv 1 \pmod 2,$$

one of the three elements $c_{ij}$, $c_{jl}$, and $c_{il}$ is $2 \pmod 4$.

By suitable modification, we may assume that

$$s = \max_{i=1,\ldots,n} \text{card}\{l : c_{il} \equiv 0 \pmod 4\}$$

and

$$c_{s+1,1} \equiv c_{s+1,2} \equiv \cdots c_{s+1,s} \equiv 0 \pmod 4.$$

Then, based on the argument in previous paragraph, we have

$$c_{ij} \equiv 2 \pmod 4$$

for all $i, j = 1, 2, \ldots, s$. In addition, we may assume that

$$t = \max_{i=1,\ldots,s} \operatorname{card}\{l: \ c_{il} \equiv 0 \pmod 4\}$$

and

$$c_{s,s+1} \equiv c_{s,s+2} \equiv \cdots \equiv c_{s,s+t} \equiv 0 \pmod 4.$$

Then we have

$$c_{ij} \equiv 2 \pmod 4$$

for $i, j = s+1, s+2, \ldots, s+t$.

If $s \geq n/2$, then we have $t \leq n-s$ and

$$\phi(C) \leq 2st + (n-s)^2 - (n-s) - t^2 + t.$$

By determining the maximum of $2st - t^2 + t$ at $0 \leq t \leq n-s$, we get

$$\phi(C) \leq 2s(n-s) \leq n^2/2.$$

If $s < n/2$, then it follows by the assumption on $s$ that

$$\phi(C) \leq n \cdot s < n^2/2.$$

As a conclusion of the two cases, the lemma is proved. □

**Proof of the second case of Theorem 3.7.** We define $C_k$ as those in the proof of the first case and assume that

$$s_k = \max_{i=1,\ldots,k} \operatorname{card}\{j: \ c_{ij} \equiv 0 \pmod 4\}$$

and

$$c_{k1} \equiv c_{k2} \equiv c_{k,s_k} \equiv 0 \pmod 4$$

in $C_k$. It is easy to see that $s_k < k$. Then, by an argument similar to the proof of the previous lemma, we can deduce

$$|c_{ij}| \geq 2$$

for $i, j = 1, 2, \ldots, s_k$ and for $i = k,\ j = s_k + 1, \ldots, k$. For convenience, we write

$$C^*_{s_n} = \begin{pmatrix} c_{s_n+1,s_n+1} & c_{s_n+1,s_n+2} & \cdots & c_{s_n+1,n} \\ c_{s_n+2,s_n+1} & c_{s_n+2,s_n+2} & \cdots & c_{s_n+2,n} \\ \vdots & \vdots & \ddots & \vdots \\ c_{n,s_n+1} & c_{n,s_n+2} & \cdots & 2 \end{pmatrix}.$$

By considering two cases $\det(S_0^*) > 0$ or $\det(S_0^*) \le 0$ and applying Lemmas 3.1 and 3.2, we get

$$\begin{aligned} \det(C) &\le (n-2)\cdot \det(C_{n-1}) + \left|\det(C_{s_n})\right| \cdot \left|\det(C^*_{s_n})\right| \\ &\le (n-2)\cdot \det(C_{n-1}) + 2(n-2)^{n-2}(n-2+2s_n). \end{aligned}$$

Repeating this process inductively, we can deduce that

$$\det(C) \le (n-2)^l \cdot \det(C_{n-l}) + 2(n-2)^{n-2}\left((n-2)l + 2\sum_{i=n-l+1}^{n} s_i\right)$$

holds for some $l$ such that $C_{n-l}$ contains no element of 0 (mod 4). Then, by Lemmas 3.2 and 3.3, we get

$$\begin{aligned} \det(C) &\le (n-2)^{n-1}(n + 2(n-l-1)) + 2(n-2)^{n-2}((n-2)l + n^2/2) \\ &= 4(n-1)^2(n-2)^{n-2}. \end{aligned}$$

The second case is proved. □

In the rest of this section we will deal with the third case of Theorem 3.7. In this case, by (3.20) we have

$$c_{ij} \equiv n \equiv 3 \pmod 4 \tag{3.21}$$

for all indices $i$ and $j$. Let $F(m, n)$ denote the family of the $m \times m$ positive definite symmetric matrices $C_m = (c_{ij})$ satisfying

$$c_{ii} = n, \quad i = 1, 2, \ldots, m$$

and

$$c_{ij} \equiv 3 \pmod 4), \ 1 \le i, j \le m,$$

and let $C^*_m$ denote a matrix in $F(m, n)$ satisfying

$$\det(C^*_m) = \max\left\{\det(C_m) : \ C_m \in F(m, n)\right\}.$$

By (3.21) it is easy to see that

$$\theta^*(n, n) \le \det(C^*_n). \tag{3.22}$$

Therefore, to prove the third case of Theorem 3.7 it is sufficient to show that

$$\det(C_n^*) \le \frac{4\cdot 11^6}{7^7}(n-3)^{n-7}n^7$$

if $n \equiv 3 \pmod 4$ and $n \ge 63$. For this purpose, first let us study the matrices $C_m^*$ in detail.

**Lemma 3.4 (Ehlich, 1964b).** *If $2 \le m \le n$, then*

$$\det(C_m^*) > (n-3)\cdot\det(C_{m-1}^*).$$

**Proof.** When $m = 2$, we may take

$$C_2^* = \begin{pmatrix} n & -1 \\ -1 & n \end{pmatrix}$$

and therefore

$$\det(C_2^*) = n^2 - 1 > (n-3)\cdot\det(C_1^*).$$

Assume that the assertion is true for $m < n$, we proceed to show it for $m+1$. If $C_m^* = (c_{ij}^*)$, we define $C_{m+1} = (c_{ij})$ by

$$c_{ij} = c_{ji} = \begin{cases} c_{ij}^* & \text{if } 1 \le i, j \le m, \\ c_{im}^* & \text{if } i \le m-1 \text{ and } j = m+1, \\ 3 & \text{if } i = m+1 \text{ and } j = m, \\ n & \text{if } i = j = m+1. \end{cases}$$

Then we have $C_{m+1} \in F_{m+1,n}$ and

$$\det(C_{m+1}) = (n-3)\cdot\det(C_m^*) + (n-3)\cdot\det(D_m),$$

where $D_m$ is different from $C_m^*$ only at the last diagonal element with $d_{mm} = 3$. On the other hand, by the inductive assumption, we have

$$\det(D_m) \ge \det(C_m^*) - (n-3)\cdot\det(C_{m-1}^*) > 0.$$

Therefore, we get

$$\det(C_{m+1}^*) \ge \det(C_{m+1}) > (n-3)\cdot\det(C_m^*).$$

The lemma is proved. □

**Lemma 3.5 (Ehlich, 1964b).** *All the non-diagonal elements of $C_m^* = (c_{ij}^*)$ are either $-1$ or 3.*

**Proof.** For convenience, let $C_{m-1}$ and $\overline{C_{m-1}}$ denote the complementary submatrices of $c_{mm}^*$ and $c_{m-1,m-1}^*$, respectively. If the lemma is false, we may assume that

$$c_{m,m-1}^* = c_{m-1,m}^* = c > 3$$

and

$$\det(C_{m-1}) \leq \det(\overline{C_{m-1}}). \tag{3.23}$$

It is clear that

$$\det(C_m^*) = (n-3)\cdot \det(C_{m-1}) + \det(D_m), \tag{3.24}$$

where $D_m$ was defined in previous lemma. By Lemma 3.4, we have

$$\begin{aligned}\det(D_m) &= \det(C_m^*) - (n-3)\cdot \det(C_{m-1}) \\ &\geq \det(C_m^*) - (n-3)\cdot \det(C_{m-1}^*) > 0.\end{aligned}$$

Then $D_m$ is positive definite and therefore

$$\det(D_m) \leq (n - \tfrac{c^2}{3})\cdot \det(\overline{D_{m-1}}) < (n-3)\cdot \det(\overline{D_{m-1}}), \tag{3.25}$$

where $\overline{D_{m-1}} = (\overline{d_{ij}})$ is the complementary submatrix of $d_{m-1,m-1}$ in $D_m$. Let $C_m = (c_{ij})$ be an $m \times m$ symmetric matrix defined by

$$c_{ij} = \begin{cases} \overline{d_{ij}} & \text{if } 1 \leq i, j \leq m-1 \text{ and } i \neq j, \\ \overline{d_{m-1,j}} & \text{if } i = m, \\ n & \text{if } i = j. \end{cases}$$

Then it follows by (3.23), (3.24), and (3.25) that

$$\begin{aligned}\det(C_m) &= (n-3)\cdot \det(\overline{C_{m-1}}) + (n-3)\cdot \det(\overline{D_{m-1}}) \\ &\geq (n-3)\cdot \det(C_{m-1}) + (n-3)\cdot \det(\overline{D_{m-1}}) \\ &> \det(C_m^*),\end{aligned}$$

which contradicts the maximum assumption on $\det(C_m^*)$. The lemma is proved. □

To characterize $C_m^*$ further, let us introduce a couple of new terms first. We say a matrix $C_m = (c_{ij}) \in F(m, n)$ has a *block* of length $l$ if, after suitable modification (exchange rows and corresponding columns)

$$c_{ij} = \begin{cases} 3 & \text{if } k \leq i, j \leq k+l-1 \text{ and } i \neq j, \\ -1 & \text{if } k \leq i \leq k+l-1 \text{ and } j < k \text{ or } j > k+l-1 \end{cases}$$

holds for some suitable $k$. If $C_m$ has $r$ blocks of lengths $l_1, l_2, \ldots, l_r$ satisfying

$$\sum_{i=1}^{r} l_i = m,$$

then we call $C_m$ a *block matrix*.

**Lemma 3.6 (Ehlich, 1964b).** *$C_m^*$ is a block matrix.*

**Proof.** If, on the contrary, $C_m^*$ is not a block matrix, then, by suitable modification, we may assume that

$$\begin{cases} c^*_{m,m-1} = c^*_{m,m-2} = 3 \\ c^*_{m-1,m-2} = -1. \end{cases}$$

Then the lemma can be proved by repeating the proof argument of Lemma 3.5. □

**Lemma 3.7 (Ehlich, 1964b).** *If $C_m \in F(m, n)$ is a block matrix with r blocks of lengths $l_1, l_2, \ldots, l_r$, respectively, then*

$$\det(C_m) = (n-3)^{m-r}\left(1 - \sum_{i=1}^{r} \frac{l_i}{n-3+4l_i}\right)\prod_{i=1}^{r}(n-3+4l_i).$$

**Proof.** When $m = 1$ or 2, the lemma is obvious. Assume that the assertion is true for $m \le k < n$, then we proceed to show it for $m = k+1$ by considering two cases.

**Case 1.** $r = k+1$. In this case it is easy to deduce that

$$\det(C_{k+1}) = (n+1)^k(n-k) = (n+1)^{k+1}\left(1 - \frac{k+1}{n+1}\right).$$

**Case 2.** $r < k+1$. For convenience, we assume that $l_r \ge 2$. Subtracting the $k$th row from the $(k+1)$th row and the $k$th column from the $(k+1)$th column we get

$$c'_{ij} = \begin{cases} 0 & \text{if } i = k+1,\ j < k \text{ or } i < k,\ j = k+1, \\ 2(n-3) & \text{if } i = j = k+1, \\ 3-n & \text{if } i = k+1,\ j = k \text{ or } i = k,\ j = k+1, \\ c_{ij} & \text{if } i, j \le k \end{cases}$$

in the resulting matrix. Therefore, we have

$$\det(C_{k+1}) = 2(n-3)\cdot\det(C_k) - (n-3)^2\cdot\det(C_{k-1}), \tag{3.26}$$

where $C_k$ and $C_{k-1}$ denote the $k \times k$ and the $(k-1)\times(k-1)$ principal submatrices of $C_{k+1}$, respectively. Clearly both $C_k$ and $C_{k-1}$ are block matrices. Now we consider two subcases.

**Subcase 2.1.** $l_r > 2$. By (3.26) and the inductive assumption, we have

$$\det(C_{k+1}) = (n-3)^{k+1-r}\Bigg[2(n-7+4l_r)\Bigg(1-\frac{l_r-1}{n-7+4l_r} - \sum_{i=1}^{r-1}\frac{l_i}{n-3+4l_i}\Bigg) - (n-11+4l_r)\Bigg(1-\frac{l_r-2}{n-11+4l_r} - \sum_{i=1}^{r-1}\frac{l_i}{n-3+4l_i}\Bigg)\Bigg]\prod_{i=1}^{r-1}(n-3+4l_i)$$
$$= (n-3)^{k+1-r}\Bigg(1-\sum_{i=1}^{r}\frac{l_i}{n-3+4l_i}\Bigg)\prod_{i=1}^{r}(n-3+4l_i).$$

**Subcase 2.2.** $l_r = 2$. Similar to the previous subcase, we get

$$\det(C_{k+1}) = (n-3)^{k+1-r}\Bigg[2(n+1)\Bigg(1-\frac{1}{n+1}-\sum_{i=1}^{r-1}\frac{l_i}{n-3+4l_i}\Bigg) -(n-3)\Bigg(1-\sum_{i=1}^{r-1}\frac{l_i}{n-3+4l_i}\Bigg)\Bigg]\prod_{i=1}^{r-1}(n-3+4l_i)$$
$$= (n-3)^{k+1-r}\Bigg(1-\sum_{i=1}^{r}\frac{l_i}{n-3+4l_i}\Bigg)\prod_{i=1}^{r}(n-3+4l_i).$$

As a conclusion of the two cases the lemma is proved. □

Writing

$$F(n) = \max\left\{(n-3)^{n-r}\Bigg(1-\sum_{i=1}^{r}\frac{l_i}{n-3+4l_i}\Bigg)\prod_{i=1}^{r}(n-3+4l_i)\right\},$$

where the maximum is over all sets of positive integers $\{l_1, l_2, \ldots, l_r\}$ satisfying $\sum_{i=1}^{r} l_i = n$, by (3.22) and Lemmas 3.5–3.7 we have

$$\theta_n^* = \theta^*(n,n) \le F(n). \tag{3.27}$$

Therefore, to prove the third case of Theorem 3.7 it is sufficient to show that

$$F(n) \le \frac{4\cdot 11^6}{7^7} n^7 (n-3)^{n-7}$$

if $n \ge 63$. For this purpose, we define

$$G(n,r) = (n-3)^{n-r}(n-3+4s)^u(n+1+4s)^v\Bigg(1-\frac{us}{n-3+4s}-\frac{v(s+1)}{n+1+4s}\Bigg),$$

where

$$\begin{cases} s = \left[\frac{n}{r}\right], \\ v = n - rs, \\ u = r - v, \end{cases} \tag{3.28}$$

and introduce a couple of technical lemmas to estimate $F(n)$.

**Lemma 3.8 (Ehlich, 1964b).**

$$F(n) = \max\left\{G(n, r) : 1 \le r \le n\right\}.$$

**Proof.** Assume that

$$F(n) = (n-3)^{n-r}\left(1 - \sum_{i=1}^{r} \frac{l_i}{n-3+4l_i}\right)\prod_{i=1}^{r}(n-3+4l_i)$$

with smallest $r$. If $r = 1$, there is nothing to prove. If $r > 1$, without loss of generality, we assume that

$$l_1 = \min\left\{l_i :\ 1 \le i \le r\right\} \quad \text{and} \quad l_2 = \max\left\{l_i :\ 1 \le i \le r\right\}.$$

We fix $l_i$ for $i \ge 3$ and assume $l_1 = x \le h/2$ and $l_2 = h - x$, where $h = l_1 + l_2$ is a constant. Then we get $F(n) = c \cdot f(x)$, where $c$ is a positive constant and

$$f(x) = (n-3+4x)(n-3+4(h-x))\left(1 - \alpha - \frac{x}{n-3+4x} - \frac{h-x}{n-3+4(h-x)}\right)$$

with

$$\alpha = \sum_{i=3}^{r} \frac{l_i}{n-3+4l_i}. \tag{3.29}$$

By a routine computation, we get

$$f'(x) = 8(h-2x)(1-2\alpha).$$

If $1 - 2\alpha \le 0$, both $f(x)$ and $F(n)$ will not decrease when we substitute $l_1$ and $l_2$ by 0 and $h$, respectively, which contradicts the assumption on $r$. If $1 - 2\alpha > 0$, both $f(x)$ and $F(n)$ will increase when we substitute $l_1$ and $l_2$ by $\left[\frac{h}{2}\right]$ and $\left[\frac{h+1}{2}\right]$, respectively. By repeating this process for at most a finite number of times we get that $u$ of the $r$ integers $\{l_1, l_2, \ldots, l_r\}$ must be $s$, and $v$ of them must be $s+1$. In other words

$$\begin{cases} s = \left[\frac{n}{r}\right], \\ v = n - rs, \\ u = r - v. \end{cases}$$

Then $F(n)$ can be expressed as some $G(n, r)$. Lemma 3.8 is proved. □

**Lemma 3.9 (Ehlich, 1964b).**

$$\max\left\{G(n,r):\ 1\le r\le n\right\}=\max\left\{G(n,r):\ 1\le r\le 7\right\}.$$

**Proof.** Assume that $r^*$ is the smallest $r$ at which the maximum is attained. We define its corresponding $f(x)$ and $\alpha$ as those in the proof of Lemma 3.8. To prove $r^*\le 7$, we proceed to show that $1-2\alpha<0$ whenever $r\ge 8$. First, let us observe two basic facts.

**Case 1.** $v=0$. If $s<(n+3)/8$, by (3.29) and (3.28), we get

$$\begin{aligned}(n-3+4s)(1-2\alpha)&=n-3+4s-2(u-2)s\\&=-n-3+8s<0.\end{aligned}$$

**Case 2.** $v>0$. If $s<(n-3)/8$, we have

$$\begin{aligned}(n+1+4s)(1-2\alpha)&\le n+1+4s-2(u-1)s-2(v-1)(s+1)\\&=-n+3+8s<0.\end{aligned}$$

In both cases, we get $1-2\alpha<0$ whenever $s<(n-3)/8$.

On the other hand, it is easy to see that, if $r\ge 12$ and $n>9$, or if $r\ge 9$ and $n>27$, we have

$$s\le\frac{n}{r}<\frac{n-3}{8}.$$

In addition, we can verify case by case that $1-2\alpha<0$ when $9\le r\le 11$ and $11\le n\le 27$. Hence, whenever $r\ge 9$ we have

$$1-2\alpha<0.$$

Now we deal with the case $r=8$. If $n\equiv 3\pmod 8$, it follows by $n=8s+3$ that

$$\begin{aligned}1-2\alpha&=1-\frac{4(s+1)}{8s+4(s+1)}-\frac{8s}{8s+4s}\\&=1-\frac{2}{3}-\frac{s+1}{3s+1}<0.\end{aligned}$$

If $n\equiv 7\pmod 8$, it follows by $n=8s+7$ that

$$1-2\alpha=1-\frac{12(s+1)}{8s+8+4s}=1-\frac{3s+3}{3s+2}<0.$$

As a conclusion of these cases, the lemma is proved. □

**Lemma 3.10 (Ehlich, 1964b).** *If $n>39$, then*

$$\max\left\{G(n,r):\ 1\le r\le 7\right\}=\max\left\{G(n,6),\ G(n,7)\right\}.$$

**Proof.** We define

$$\beta = \frac{(u-1)s}{n-3+4s} + \frac{v(s+1)}{n+1+4s}$$

and

$$G(n, r, x) = (n-3)^{n-r-1}(n-3+4s)^{u-1}(n+1+4s)^{v} g(x),$$

where

$$g(x) = (n-3+4x)(n-3+4s-4x)\left(1-\beta-\frac{x}{n-3+4x}-\frac{s-x}{n-3+4s-4x}\right).$$

By a routine computation, we get

$$G(n, r) = G(n, r, 0) = G(n, r, s).$$

In fact, both $G(n, r, x)$ and $g(x)$ attain their maxima at $x = 0$. Otherwise, we can easily deduce $G(n, r, x) > G(n, r, 0)$ for $0 < x < s$ and hence can construct a block matrix of larger determinant. Since

$$g'(x) = 8(s-2x)(1-2\beta),$$

we must have $1-2\beta < 0$. However, if $r \leq 5$ and $n > 39$, by (3.28) and the definition of $\beta$ we have

$$s \geq \frac{n-4}{5} > \frac{n+3}{6},$$

$$(n-3+4s)(1-2\beta) \geq -n-3+6s > 0,$$

and therefore

$$1-2\beta > 0.$$

By this contradiction, the lemma is proved. □

**Proof of the third case of Theorem 3.7.** By (3.27) and Lemmas 3.8–3.10, it is sufficient to show that

$$G(n, 6) \leq G(n, 7) \leq \frac{4 \cdot 11^6}{7^7} n^7 (n-3)^{n-7} \tag{3.30}$$

for $n \geq 63$ and $n \equiv 3 \pmod 4$.

It is easy to see that

$$1 - \frac{n}{n-3+4s} \leq 1 - \frac{us}{n-3+4s} - \frac{v(s+1)}{n+1+4s} \leq 1 - \frac{n}{n+1+4s}.$$

Therefore, we have

$$(n-3)^{r-n} G(n, r) \geq (n-3+4s)^{u-1}(n+1+4s)^{v}(4s-3) \tag{3.31}$$

and for $v > 0$

$$(n-3)^{r-n}G(n,r) < (n-3+4s)^u(n+1+4s)^{v-1}(4s+1). \tag{3.32}$$

When $n \geq 63$, we have

$$(11n-41)(11n-17) \leq (11n-21)^2$$

and

$$(11n-17)(4n-45) \geq (11n-33)(4n-40).$$

Therefore, by considering different $\{u, v\}$ pairs in (3.31), it follows that

$$\begin{aligned} 7^7(n-3)^{7-n}G(n,7) &\geq (11n-41)^5(11n-17)(4n-45) \\ &\geq 4\cdot 11^6(n-4)^5(n-10)(n-3). \end{aligned}$$

Similarly, it follows by (3.32) that

$$\begin{aligned} 3^6(n-3)^{6-n}G(n,6) &< 2(5n+1)^5(n-1) \\ &< 2\cdot 5^5(n+1)^5(n-1). \end{aligned}$$

Hence for $n > 10^4$, we have

$$\begin{aligned} \frac{G(n,7)}{G(n,6)} &> \frac{2\cdot 3^6\cdot 11^6}{5^5\cdot 7^7}\left(1-\frac{5}{n+1}\right)^5\left(1-\frac{9}{n-1}\right) \\ &> 1.00363\cdot 0.9995^5\cdot 0.999 \\ &> 1.000117. \end{aligned}$$

In addition, it can be verified by a computer that

$$\max\left\{G(n,6),\ G(n,7)\right\} = G(n,7)$$

when $63 \leq n \leq 10^4$ and $n \equiv 3 \pmod 4$.

Finally, by considering different $\{u, v\}$ pairs in (3.32), we get

$$G(n,7) \leq \frac{4\cdot 11^6}{7^7}n^7(n-3)^{n-7}.$$

By (3.30) the third case of the theorem is proved. □

**Remark 3.5.** It is natural to ask: Are the upper bounds in Theorem 3.7 tight? To answer this question, by some skilful constructions, Neubauer and Radcliffe (1997) were able to show that both the first two bounds can be attained at an infinite number of $n$. In the third case, we are not so lucky.

Based on Theorem 3.3 and Remark 3.4, Theorem 3.6 and Theorem 3.7 together can be restated in terms of inscribed simplices in $I^n$ as follows.

**Theorem 3.7***. *Let $S$ denote an $n$-dimensional simplex contained in $I^n$. Then*

$$v_n(S) \leq \begin{cases} \frac{1}{n!2^n}\sqrt{(2n+1)n^n} & \text{if } n \equiv 0 \pmod 4, \\ \frac{1}{(n-1)!2^{n-1}}\sqrt{(n-1)^{n-1}} & \text{if } n \equiv 1 \pmod 4, \\ \frac{11^3}{n!2^{n-1}}\sqrt{\frac{(n-2)^{n-6}(n+1)^7}{7^7}} & \text{if } n \equiv 2 \pmod 4 \text{ and } n \geq 62, \\ \frac{1}{n!2^n}\sqrt{(n+1)^{n+1}} & \text{if } n \equiv 3 \pmod 4. \end{cases}$$

We end this section with the following problem.

**Problem 3.4.** Determine the value of

$$\gamma = \liminf_{n\to\infty} \frac{\theta^*(n, n)}{n^n}.$$

Is the sequence $\{\theta^*(n, n)/n^n : n = 1, 2, \ldots\}$ dense in $[\gamma, 1]$?

## 3.4 Some particular cases

In this section we deal with the following problems.

1 What is the maximum area of a triangle inscribed in $I^n$?
2 What is the maximum volume of a tetrahedron inscribed in $I^n$?
3 For relatively small $n$, what do we know about the maximum volume of the $n$-dimensional simplex inscribed in $I^n$?

Let $\gamma(n, k)$ denote the maximum volume of a $k$-dimensional simplex inscribed in $I^n$, especially let us abbreviate $\gamma(n, n)$ to $\gamma_n$. By Theorem 3.3 and the definition of $\theta(n, k)$, we have

$$\gamma(n, k) = \frac{1}{k!}\sqrt{\theta(n, k)}. \tag{3.33}$$

Now we can determine $\gamma(n, 2)$ and $\gamma(n, 3)$ by studying binary matrices.

**Theorem 3.8 (Hudelson, Klee, and Larman, 1996, and Neubauer, Watkins, and Zeitlin, 1997).** *Write $l = [n/3]$ and $j = n - 3l$, then*

$$\gamma(n, 2) = \begin{cases} \frac{1}{2}\sqrt{3l^2} & \text{if } j = 0, \\ \frac{1}{2}\sqrt{3l^2 + 2l} & \text{if } j = 1, \\ \frac{1}{2}\sqrt{3l^2 + 4l + 1} & \text{if } j = 2. \end{cases}$$

**Proof.** If $A$ is a $2 \times n$ binary matrix with $l_1$ columns identical with $(1, 0)'$, $l_2$ columns identical with $(0, 1)'$, and $l_3$ columns identical with $(1, 1)'$, then we have

$$AA' = \begin{pmatrix} l_1+l_3 & l_3 \\ l_3 & l_2+l_3 \end{pmatrix}$$

and therefore

$$\det(AA') = f(l_1, l_2, l_3) = l_1l_2 + l_1l_3 + l_2l_3.$$

Since $l_1+l_2+l_3 = n$ and $l_i \geq 0$, assuming that $l_1 \leq l_2 \leq l_3$, it is easy to show that $f(l_1, l_2, l_3)$ attains its maximum if and only if $l_1$, $l_2$, and $l_3$ are balanced; that is, $l_3 - l_1 \leq 1$. Consequently, $l_1$ must be $l$. Then, by considering three cases, we get

$$f(l_1, l_2, l_3) \leq \begin{cases} 3l^2 & \text{if } j=0, \\ 3l^2+2l & \text{if } j=1, \\ 3l^2+4l+1 & \text{if } j=2, \end{cases}$$

where the upper bounds can be attained in every case. By (3.33) the theorem is proved. □

**Theorem 3.9 (Hudelson, Klee, and Larman, 1996, and Neubauer, Watkins, and Zeitlin, 1997).** *Write* $l = [n/3]$ *and* $j = n - 3l$, *then*

$$\gamma(n, 3) = \tfrac{1}{3}\sqrt{l^{3-j}(l+1)^j}.$$

**Proof.** Let $S = \text{conv}\{\mathbf{o}, \mathbf{a}_1, \mathbf{a}_2, \mathbf{a}_3\}$ be a tetrahedron inscribed in $\overline{I^n}$ and let $A = (a_{ij})$ be the corresponding $3 \times n$ binary matrix. If three of the four vertices of $S$ belong to a facet of $\overline{I^n}$, then it follows by Theorem 3.8 that

$$v_3(S) \leq \tfrac{1}{3}\gamma(n-1, 2) < \tfrac{1}{3}\sqrt{l^{3-j}(l+1)^j}.$$

Therefore, if

$$v_3(S) \geq \tfrac{1}{3}\sqrt{l^{3-j}(l+1)^j},$$

since $\mathbf{o} = (0, 0, \ldots, 0)$ is a vertex of $S$, $A$ has no column identical with $(1, 0, 0)'$, $(0, 1, 0)'$, $(0, 0, 1)'$, or $(1, 1, 1)'$. Assume that $A$ has $l_1$ columns identical with $(1, 1, 0)'$, $l_2$ columns identical with $(1, 0, 1)'$, and $l_3$ columns identical with $(0, 1, 1)'$, then we have

$$l_1 + l_2 + l_3 = n$$

and

$$AA' = \begin{pmatrix} l_1+l_2 & l_1 & l_2 \\ l_1 & l_1+l_3 & l_3 \\ l_2 & l_3 & l_2+l_3 \end{pmatrix}.$$

Therefore we get

$$\det(AA') = 4l_1 l_2 l_3 \le 4l^{3-j}(l+1)^j,$$

where equality holds if and only if $l_1$, $l_2$, and $l_3$ are balanced; that is

$$\max\{l_1, l_2, l_3\} \le \min\{l_1, l_2, l_3\} + 1.$$

Consequently, we get

$$\gamma(n,3) = \tfrac{1}{3}\sqrt{l^{3-j}(l+1)^j}.$$

The theorem is proved. □

**Remark 3.6.** By studying D-optimal designs, Neubauer, Watkins, and Zeitlin (1998b, 2000) were able to determine $\gamma(n,4)$, $\gamma(n,5)$, and $\gamma(n,6)$. For sufficiently large $n$, Neubauer and Watkins (2002) obtained $\gamma(n,7)$ and Ábrego, Fernández–Merchant, Neubauer, and Watkins (2003) determined $\gamma(n,11)$, $\gamma(n,15)$, $\gamma(n,19)$, and $\gamma(n,23)$.

Clearly, among the problems concerning inscribed simplices in $I^n$, to determine the exact value of $\gamma_n$ for a particular $n$ is both important and interesting. As mentioned in Remark 3.4 and Remark 3.5, we do know the value of $\gamma_n$ for an infinite number of $n$ (see Agaian, 1985; Neubauer and Radcliffe, 1997). To end this chapter, let us list the known results about $\theta_n$, $\theta_n^*$ (defined in Remark 3.2), and $\gamma_n$ up to $n = 11$ as follows.

| n | 2 | 3 | 4 | 5 | 6 |
|---|---|---|---|---|---|
| $\theta_{n+1}^*$ | 4 | 16 | 48 | 160 | 576 |
| $\theta_n = \frac{\theta_{n+1}^*}{2^n}$ | 1 | 2 | 3 | 5 | 9 |
| $\gamma_n = \frac{\theta_n}{n!}$ | 0.5 | 0.3333333 | 0.125 | 0.0416666 | 0.0125 |
| Author | Williamson | Sylvester | Ehlich | Ehlich | Williamson |

| | | | | |
|---|---|---|---|---|
| 7 | 8 | 9 | 10 | 11 |
| 4096 | 14336 | 73728 | 327680 | 2985984 |
| 32 | 56 | 144 | 320 | 1458 |
| 0.00063492 | 0.0013888 | 0.0003968 | 0.0000881 | 0.0000365 |
| Sylvester | Ehlich and Zeller | Ehlich | Ehlich | Hadamard |

# 4
# Triangulations

## 4.1 An example

Let $\mathbf{v}_1, \mathbf{v}_2, \ldots, \mathbf{v}_8$ be the eight vertices of the unit box $I^3$, as illustrated in Figure 4.1. The box can be divided into six tetrahedra $T_1, T_2, \ldots, T_6$ defined by

$$T_1 = \text{conv}\{\mathbf{v}_1, \mathbf{v}_5, \mathbf{v}_6, \mathbf{v}_8\}, \qquad T_2 = \text{conv}\{\mathbf{v}_1, \mathbf{v}_6, \mathbf{v}_7, \mathbf{v}_8\},$$

$$T_3 = \text{conv}\{\mathbf{v}_1, \mathbf{v}_2, \mathbf{v}_6, \mathbf{v}_7\}, \qquad T_4 = \text{conv}\{\mathbf{v}_1, \mathbf{v}_2, \mathbf{v}_3, \mathbf{v}_7\},$$

$$T_5 = \text{conv}\{\mathbf{v}_1, \mathbf{v}_3, \mathbf{v}_4, \mathbf{v}_8\}, \quad \text{and} \quad T_6 = \text{conv}\{\mathbf{v}_1, \mathbf{v}_3, \mathbf{v}_7, \mathbf{v}_8\}.$$

Then we have

$$v_3(T_i) = \tfrac{1}{3!}$$

for all $i = 1, 2, \ldots, 6$.

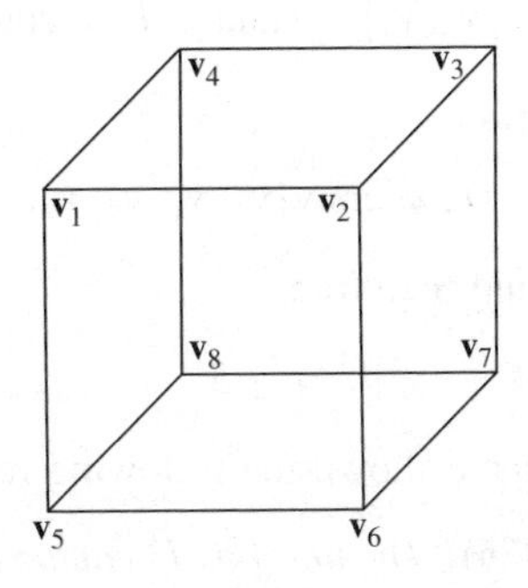


Figure 4.1

Can we do better? In other words, can we divide the box into a set of tetrahedra of smaller cardinality? Yes, in fact, we can divide $I^3$ into five tetrahedra $T'_1, T'_2, \ldots, T'_5$ defined by

$$T'_1 = \mathrm{conv}\{\mathbf{v}_1, \mathbf{v}_2, \mathbf{v}_3, \mathbf{v}_6\}, \qquad T'_2 = \mathrm{conv}\{\mathbf{v}_1, \mathbf{v}_3, \mathbf{v}_4, \mathbf{v}_8\},$$

$$T'_3 = \mathrm{conv}\{\mathbf{v}_1, \mathbf{v}_5, \mathbf{v}_6, \mathbf{v}_8\}, \qquad T'_4 = \mathrm{conv}\{\mathbf{v}_3, \mathbf{v}_6, \mathbf{v}_7, \mathbf{v}_8\},$$

and

$$T'_5 = \mathrm{conv}\{\mathbf{v}_1, \mathbf{v}_3, \mathbf{v}_6, \mathbf{v}_8\}.$$

In this case, we can verify that

$$v_3(T'_i) = \tfrac{1}{3!}$$

for all $i = 1$, 2, 3, and 4, and

$$v_3(T'_5) = \tfrac{2}{3!}.$$

Can we do still better? The answer is "no." Assume that the unit box $I^3$ can be divided into $\tau$ tetrahedra $T_1, T_2, \ldots, T_\tau$ and its facets are divided into $t$ corresponding triangles. Then, since $I^3$ has six facets, we have

$$t \geq 12. \tag{4.1}$$

It is easy to see that each of the triangles belongs to one tetrahedron and each of the tetrahedra takes at most three of the triangles as its facets. Thus we have

$$\tau \geq \tfrac{12}{3} = 4. \tag{4.2}$$

If equality holds in (4.2), then equality in (4.1) must be attained as well. Therefore, after suitable permutation, we have

$$T_1 = \mathrm{conv}\{\mathbf{v}_1, \mathbf{v}_2, \mathbf{v}_4, \mathbf{v}_5\}, \qquad T_2 = \mathrm{conv}\{\mathbf{v}_2, \mathbf{v}_3, \mathbf{v}_4, \mathbf{v}_7\},$$

$$T_3 = \mathrm{conv}\{\mathbf{v}_2, \mathbf{v}_5, \mathbf{v}_6, \mathbf{v}_7\}, \quad \text{and} \quad T_4 = \mathrm{conv}\{\mathbf{v}_4, \mathbf{v}_5, \mathbf{v}_7, \mathbf{v}_8\}.$$

Then a fifth tetrahedron

$$T_5 = \mathrm{conv}\{\mathbf{v}_2, \mathbf{v}_4, \mathbf{v}_5, \mathbf{v}_7\}$$

must be on the list too and therefore

$$\tau \geq 5.$$

As a conclusion, we have shown the following result.

**Theorem 4.1 (Mara, 1976).** *The unit box $I^3$ can be divided into 5 tetrahedra, but not fewer.*

## 4.2 Some special triangulations

Let $V(P)$ denote the set of the vertices of a polytope $P$. A finite set $\wp$ of $n$-dimensional polytopes is called an *n-complex* if the intersection of every pair of the polytopes is a common face of them. An $n$-complex $\Im$ of simplices is called a *triangulation*[4] of $\wp$ if

$$\bigcup_{S\in\Im} S = \bigcup_{P\in\wp} P,$$

for every $P \in \wp$ there is a subset $\Im^* \subseteq \Im$ such that

$$P = \bigcup_{S\in\Im^*} S$$

and, for every $S \in \Im^*$

$$V(S) \subseteq V(P). \tag{4.3}$$

Clearly, for a triangulation $\Im$ of $\wp$, its cardinality is a natural measure of efficiency. We define

$$\tau(\wp) = \min_{\Im} \left\{\mathrm{card}(\Im)\right\},$$

where the minimum is over all the triangulations $\Im$ of $\wp$. To find efficient triangulations and to determine the value of $\tau(\wp)$ for general $n$-complexes is important, fascinating, and challenging. In this chapter, we only focus on the very special case $\wp = I^n$. For convenience, we abbreviate $\tau(I^n)$ to $\tau_n$.

In this section, we will introduce several special triangulations for $I^n$ and therefore produce some corresponding upper bounds for $\tau_n$.

**Triangulation I.** Assume that $\Im = \{S_1, S_2, \ldots, S_k\}$ is a triangulation of $I^n$. By Theorem 3.3, it follows that

$$v_n(S_i) = \frac{m_i}{n!}$$

holds for some suitable positive integer $m_i$. Therefore, by comparing these volumes with the volume of $I^n$, we have

$$\tau_n \le n!. \tag{4.4}$$

In fact, $I^n$ indeed has a triangulation of cardinality $n!$. The two-dimensional case is obvious. It is known that $I^n$ has $n$ facets $F_1, F_2, \ldots, F_n$ opposite to a given vertex $\mathbf{v}$, and all the facets are $(n-1)$-dimensional unit cubes. Assume that $\{S_{i,j} : j = 1, 2, \ldots, (n-1)!\}$ is a triangulation for $F_i$, then

[4] Triangulation is a particular type of decomposition, which will be introduced in Section 4.4.

$\{\text{conv}(\mathbf{v} \cup S_{i,j}) : i = 1, 2, \dots, n;\ j = 1, 2, \dots, (n-1)!\}$ will be a triangulation of cardinality $n!$ for $I^n$.

**Triangulation II (Sallee, 1982a, 1982b, and 1984).** If $P = F_n \supset F_{n-1} \supset \cdots \supset F_0 \neq \emptyset$ is a sequence of faces of a polytope $P$ and $\mathbf{v}_n, \mathbf{v}_{n-1}, \dots, \mathbf{v}_0$ is a sequence of vertices of $P$ such that $\mathbf{v}_i \in F_i$ and $\mathbf{v}_{i+1} \notin F_i$ hold for $0 \leq i \leq n-1$, then $\text{conv}\{\mathbf{v}_0, \mathbf{v}_1, \dots, \mathbf{v}_n\}$ is an $n$-dimensional simplex. Let $V = \{\mathbf{v}_1, \mathbf{v}_2, \dots, \mathbf{v}_m\}$ be an ordering of the vertices of an $n$-complex $\wp$. For each face $F$ of $P \in \wp$, we define

$$i(F) = \min\left\{i : \mathbf{v}_i \in F\right\}$$

and

$$\mathbf{v}(F) = \mathbf{v}_{i(F)}.$$

Then each sequence of faces $P = F_n \supset F_{n-1} \supset \cdots \supset F_0 \neq \emptyset$ satisfying $\mathbf{v}(F_{i+1}) \notin F_i$ for $0 \leq i \leq n-1$ has an associated simplex $\text{conv}\{\mathbf{v}(F_n), \mathbf{v}(F_{n-1}), \dots, \mathbf{v}(F_0)\}$. Let $\Im(\wp, V)$ denote the set of all simplices generated by sequences of faces defined as above. Then we have the following lemma.

**Lemma 4.1 (Sallee, 1982a).** *Let $\wp$ be a complex with a vertex ordering $V$. Then the corresponding $\Im(\wp, V)$ is a triangulation of $\wp$.*

**Proof.** For $n = 2$ the assertion is obvious. Assuming that the statement is true for $n = k$, we proceed to show it for $n = k + 1$.

Let $\wp$ be a $(k+1)$-complex and let $\wp'$ denote its facet complex, the $k$-complex consisting of the facets of the polytopes of $\wp$. By the assumption $\Im(\wp', V)$ is a triangulation of $\wp'$, then, for each polytope $P \in \wp$, the $k$-complex of facets opposite $\mathbf{v}(P)$ is triangulated by a subset $\Im^*$ of $\Im(\wp', V)$. Thus

$$\Im(P) = \left\{\text{conv}\left\{\mathbf{v}(P) \cup S\right\} : S \in \Im^*\right\}$$

is a triangulation of $P$.

On the other hand, each simplex in $\Im^*$ has an associated sequence of faces $F_k \supset F_{k-1} \supset \cdots \supset F_0 \neq \emptyset$. Since $\mathbf{v}(P) \notin F_k$, this sequence can be extended into a sequence $P = F_{k+1} \supset F_k \supset \cdots \supset F_0 \neq \emptyset$. Thus

$$\Im(\wp, V) = \bigcup_{P \in \wp} \Im(P)$$

is a triangulation of $\wp$. The lemma is proved. □

Based on this lemma, we can construct some relatively efficient triangulations for $\overline{I^n}$. For example, we define

$$P_1(n,k) = \left\{\mathbf{x} \in \overline{I^n} : \sum x_i \leq k\right\},$$

$$P_2(n,k) = \left\{\mathbf{x} \in \overline{I^n} : \sum x_i \geq k\right\},$$

and

$$\wp(n,k) = \Big\{P_1(n,k),\ P_2(n,k)\Big\}.$$

It is easy to see that

$$\overline{I^n} = P_1(n,k) \bigcup P_2(n,k).$$

In addition, since the two endpoints of an edge of $\overline{I^n}$ are different only at one coordinate and the difference is one, we have

$$V(P_i(n,k)) \subseteq V(\overline{I^n}).$$

Let $V$ be an ordering of $V(\overline{I^n})$ satisfying the following conditions.

1 *If* $\sum v_i < k$ *and* $\sum v_i < \sum u_i$, *then* $\mathbf{v}$ *is after* $\mathbf{u}$.
2 *If* $k < \sum v_i < \sum u_i$, *then* $\mathbf{v}$ *is before* $\mathbf{u}$.

By Lemma 4.1, the complex $\Im(\wp(n,k), V)$ does provide a triangulation for $\wp(n,k)$ and consequently for $\overline{I^n}$. For the special case $k = [n/2]$, based on some detailed combinatorial analysis, Sallee (1984) deduced that

$$\tau_n \leq \mathrm{card}\Big\{\Im(\wp(n,[n/2]), V)\Big\} \leq o(1) \cdot n!. \tag{4.5}$$

**Triangulation III (Haiman, 1991).** For convenience, let $S^i$ denote an $i$-dimensional simplex. In 1988, Billera, Cushman, and Sanders proved the following result about the triangulations of a Cartesian product of simplices.

**Lemma 4.2.** *Every triangulation of* $S^k \oplus S^l$ *has exactly* $(k+l)!/k! \cdot l!$ *simplices.*

**Proof.** Let $S$ be a $(k+l)$-dimensional vertex simplex of $S^k \oplus S^l$. We proceed to show that

$$\frac{v(S)}{v(S^k \oplus S^l)} = \frac{k! \cdot l!}{(k+l)!}. \tag{4.6}$$

Since the ratio $v(S)/v(S^k \oplus S^l)$ keeps invariant under nonsingular affine transformations, without loss of generality we assume that $S^k = \mathrm{conv}\{\mathbf{o}, \mathbf{e}_1, \ldots, \mathbf{e}_k\}$, $S^l = \mathrm{conv}\{\mathbf{o}, \mathbf{e}_{k+1}, \ldots, \mathbf{e}_{k+l}\}$, and $S = \mathrm{conv}\{\mathbf{o}, \mathbf{a}_1, \ldots, \mathbf{a}_{k+l}\}$. Then it is easy to see that $S^k \oplus S^l$ is a $(k+l)$-dimensional polytope with vertices $\mathbf{o}$, $\mathbf{e}_i$, and $\mathbf{e}_{j_1} \oplus \mathbf{e}_{j_2}$, where $1 \le i \le k+l$, $1 \le j_1 \le k$ and $k+1 \le j_2 \le k+l$. Thus we have

$$v(S^k \oplus S^l) = v_k(S^k) \cdot v_l(S^l) = \frac{1}{k! \cdot l!} \tag{4.7}$$

and

$$v(S) = \frac{|\det(A)|}{(k+l)!}, \tag{4.8}$$

where $A$ is the $(k+l) \times (k+l)$ matrix with entries $a_{ij}$.

If $\mathbf{a}_i = \mathbf{e}_j$ holds for some $i$ and $j$, then by induction we can show

$$|\det(A)| = 1. \tag{4.9}$$

In this case, (4.6) follows from (4.7), (4.8), and (4.9). If, on the contrary

$$\mathbf{a}_i = \mathbf{e}_{j_1} \oplus \mathbf{e}_{j_2} \tag{4.10}$$

holds for all indices $i$, then we claim that

$$\det(A) = 0,$$

which contradicts the assumption that $S$ is $(k+l)$-dimensional. To see this, we may look at a $(k, l)$-*bipartite graph* $G$ with $k+l$ vertices $\mathbf{v}_1, \mathbf{v}_2, \ldots, \mathbf{v}_{k+l}$ such that $\mathbf{v}_{j_1}$ and $\mathbf{v}_{j_2}$ is connected if and only if $\mathbf{e}_{j_1} \oplus \mathbf{e}_{j_2}$ is a vertex of $S$. Since $G$ is a bipartite graph with $k+l$ vertices, it can be easily shown that each set of $k+l$ edges contains an even cycle. Without loss of generality, assume that $\{\mathbf{a}_1, \mathbf{a}_2, \ldots, \mathbf{a}_{2m}\}$ is an even cycle in $\{\mathbf{a}_1, \mathbf{a}_2, \ldots, \mathbf{a}_{k+l}\}$, then by (4.10) we get

$$(-1)^1\mathbf{a}_1 + (-1)^2\mathbf{a}_2 + \cdots + (-1)^{2m}\mathbf{a}_{2m} = 0$$

and hence

$$\det(A) = 0.$$

The lemma is proved. □

Since $I^{k+l} = I^k \oplus I^l$, based on Lemma 4.2, Haiman (1991) discovered the following triangulation for $I^{k+l}$. If $\Im_k = \{S_1^k, S_2^k, \ldots, S_{\tau_k}^k\}$ is a triangulation for $I^k$ and $\Im_l = \{S_1^l, S_2^l, \ldots, S_{\tau_l}^l\}$ is a triangulation for $I^l$, then

$$\wp = \left\{ S_i^k \oplus S_j^l : \; i = 1, 2, \ldots, \tau_k; \; j = 1, 2, \ldots, \tau_l \right\}$$

is a $(k+l)$-complex satisfying

$$I^{k+l} = \bigcup_{i,j} S_i^k \oplus S_j^l.$$

By Lemma 4.1, we can produce a triangulation

$$\Im = \bigcup_{i,j} \Im_{i,j}$$

for $\wp$, where $\Im_{i,j}$ is a triangulation of $S_i^k \oplus S_j^l$. By Lemma 4.2, we have

$$\mathrm{card}\{\Im\} = \frac{(k+l)! \cdot \tau_k \cdot \tau_l}{k! \cdot l!}.$$

Consequently, for any fixed $k$, we can deduce

$$\tau_{nk} \leq \frac{(nk)! \cdot \tau_{(n-1)k} \cdot \tau_k}{((n-1)k)! \cdot k!} \leq \left(\frac{\tau_k}{k!}\right)^n \cdot (nk)! = \rho_k^{nk} \cdot (nk)!, \tag{4.11}$$

where

$$\rho_k = \sqrt[k]{\frac{\tau_k}{k!}}.$$

Especially, since $\tau_k$ is known for some small $k$ (see Section 4.4), we can deduce the following upper bound from (4.11).

**Theorem 4.2 (Haiman, 1991).** *When $n$ is large*

$$\tau_n \leq 0.871^n \cdot n!.$$

**Remark 4.1.** Recently, Orden and Santos (2003) were able to reduce 0.871 to 0.816. So far this is the best-known upper bound for $\tau_n$.

## 4.3 Smith's lower bound

Let $\Im = \{S_1, S_2, \ldots, S_{\tau_n}\}$ be a triangulation for $I^n$. By Theorem 3.5*, we have

$$v_n(S_i) \leq \frac{(n+1)^{\frac{n+1}{2}}}{2^n n!}$$

and therefore

$$\tau_n \geq \frac{1}{\max\{v_n(S_i)\}} \geq \frac{2^n n!}{(n+1)^{\frac{n+1}{2}}}. \tag{4.12}$$

This is the first lower bound for $\tau_n$. For the convenience of further comparison, we abbreviate the number on the right-hand side of (4.12) to $\eta_n$. By *Stirling's formula*, we get

$$\eta_n \sim \sqrt{\frac{2\pi}{e}} \left(\frac{2}{e}\right)^n n^{\frac{n}{2}}.$$

Let $F$ be a facet of $I^n$ and let $\Im$ be a triangulation of $I^n$. We will say that a simplex $S \in \Im$ belongs to $F$ if a facet of $S$ is a subset of $F$. If $\Im_1$ and $\Im_2$

are the sets of simplices belonging to two adjacent facets of $I^n$, respectively, then it is easy to see that

$$\sum_{S\in\Im_1} v_n(S) = \frac{1}{n} \tag{4.13}$$

and

$$\sum_{S\in\Im_1\cap\Im_2} v_n(S) \leq \frac{1}{n(n-1)}. \tag{4.14}$$

Let $\{F_i, -F_i\}$, $i = 1, 2, \ldots, n$, be the $n$ distinct pairs of opposite facets of $I^n$ and define

$$\aleph_i = \Big\{ S \in \Im :\ S \text{ belongs to } F_i \text{ or } -F_i \Big\}$$

and

$$\Re_k = \bigcup_{i=1}^{k} \aleph_i.$$

By a combinatorial argument and (4.14), it follows that

$$\sum_{S\in\Re_{k-1}\cap\aleph_k} v_n(S) \leq \frac{4(k-1)}{n(n-1)}.$$

Hence, by induction and applying (4.13) and (4.14), we get

$$\begin{aligned}\sum_{S\in\Re_k} v_n(S) &= \sum_{S\in\Re_{k-1}} v_n(S) + \sum_{S\in\aleph_k} v_n(S) - \sum_{S\in\Re_{k-1}\cap\aleph_k} v_n(S)\\ &\geq \sum_{i=1}^{k} \left(\frac{2}{n} - \frac{4(i-1)}{n(n-1)}\right)\end{aligned}$$

and especially

$$\sum_{S\in\Re_{[(n+1)/2]}} v_n(S) \geq \tfrac{1}{2}.$$

On the other hand, for all $S \in \Re_{[(n+1)/2]}$, we have

$$v_n(S) \leq \frac{1}{n\,\eta_{n-1}}.$$

Thus we get

$$\tau_n \geq \tfrac{1}{2}\left(\eta_n + n\,\eta_{n-1}\right) \gg \left(\tfrac{1}{2} + \tfrac{1}{4}\sqrt{e(n+1)}\right)\eta_n, \tag{4.15}$$

which improves (4.12) by an asymptotic factor $0.25\sqrt{en}$. This lower bound was discovered by Sallee (1984).

Note that both (4.12) and (4.15) are based on volume estimation of the simplices, to get a better lower bound for $\tau_n$ it is natural to introduce a weight function to the measure. In fact, this idea can be realized by applying some basic results in Hyperbolic Geometry. By such an approach, W.D. Smith was able to prove the following theorem.

**Theorem 4.3 (Smith, 2000).**

$$\tau_n \geq \tfrac{1}{2}\left(\tfrac{3}{2}\right)^{\frac{n}{2}} \eta_n.$$

The $n$-dimensional *hyperbolic space* has several models. Its *Poincaré disk model* $H^n$ is defined as the open unit ball $\mathrm{int}(B^n)$ with the *Riemannian metric*

$$ds^2 = \frac{4}{(1-\|\mathbf{x}\|^2)^2} \sum_{i=1}^{n} (dx_i)^2.$$

In this model the *geodesics* are circles orthogonal to the boundary of the unit ball and an $n$-dimensional $h$-simplex with vertices $\mathbf{v}_0, \mathbf{v}_1, \ldots, \mathbf{v}_n \in B^n$ is the closed subset of $H^n$ bounded by the $n+1$ spheres, which contain all the vertices except one and which are orthogonal to the boundary of the unit ball.

Let $H_p^n$ denote the *projective model* of the $n$-dimensional hyperbolic space; that is, the open ball $\mathrm{int}(B^n)$ with the Riemannian metric

$$ds^2 = \frac{1}{1-\|\mathbf{x}\|^2} \sum_{i=1}^{n} dx_i^2 + \frac{1}{(1-\|\mathbf{x}\|^2)^2} \sum_{i,j} x_i x_j \, dx_i dx_j$$

and the volume form

$$d\varpi = (1-\|\mathbf{x}\|^2)^{-\frac{n+1}{2}} d\mathbf{x}. \tag{4.16}$$

In fact, we can obtain $H_p^n$ from $H^n$ by a map

$$g_1(\mathbf{x}) = \frac{2}{1+\|\mathbf{x}\|^2} \mathbf{x}.$$

The advantage of the projective model is that the geodesics become straight lines in $B^n$. Therefore, if $S$ is an $h$-simplex in $H^n$ with vertices on $\partial(B^n)$, then $g_1(S)$ is simply the Euclidean simplex with the same vertices.

Let $H_s^n$ denote the *half space model* of the $n$-dimensional hyperbolic space; that is, the half space $\{\mathbf{x} \in R^n : x_n > 0\}$ with the Riemannian metric

$$ds^2 = \frac{1}{x_n^2} \sum_{i=1}^{n} (dx_i)^2$$

and the volume form

$$d\varpi = x_n^{-n} d\mathbf{x}. \tag{4.17}$$

In fact, we can obtain $H_s^n$ from $H^n$ by a map

$$g_2(\mathbf{x}) = \frac{1}{\|\mathbf{x}-\mathbf{e}_n\|^2}\Big(2x_1, 2x_2, \ldots, 2x_{n-1}, 1-\|\mathbf{x}\|^2\Big).$$

The geodesics in $H_s^n$ are half circles and half lines orthogonal to the plane $P = \{\mathbf{x} : x_n = 0\}$.

Let $S^n$ be a maximal simplex contained in $H_p^n$ and let $\overline{S^n}$ be a maximal regular simplex contained in $H_p^n$, both with respect to $\varpi$. It is easy to see that the vertices of both $S^n$ and $\overline{S^n}$ are on the boundary of $H_p^n$. Assume that $S = \mathrm{conv}\{\mathbf{v}_0, \mathbf{v}_1, \ldots, \mathbf{v}_n\}$ is a simplex with $\mathbf{v}_n = \mathbf{e}_n$ and define

$$S' = \mathrm{conv}\Big\{g(\mathbf{v}_0), g(\mathbf{v}_1), \ldots, g(\mathbf{v}_{n-1})\Big\},$$

where $g(\mathbf{x}) = g_2(g_1^{-1}(\mathbf{x}))$ and $\mathbf{x}' = (x_1, x_2, \ldots, x_{n-1})$. Then, by (4.16) and (4.17), we have

$$\varpi(S) = \int_S (1-\|\mathbf{x}\|^2)^{-\frac{n+1}{2}}\, d\mathbf{x} \tag{4.18}$$

and

$$\begin{aligned}\varpi(S) &= \int_{S'} \left(\int_{(1-\|\mathbf{x}'\|^2)^{1/2}}^{\infty} t^{-n} dt\right) d\mathbf{x}' \\ &= \frac{1}{n-1}\int_{S'} (1-\|\mathbf{x}'\|^2)^{-\frac{n-1}{2}}\, d\mathbf{x}'. \end{aligned} \tag{4.19}$$

It is well known that the following actions do not change the measure of a set in $H_s^n$: translations parallel to $P$, rotations leaving the $x_n$-axis pointwise fixed, and multiplications by positive scalars. Therefore, we may assume that $S'$ is inscribed in $B^{n-1}$. Especially, we can see that $S'$ will be regular if $S$ is regular. For convenience, we write

$$\varphi(n, z) = \int_{\overline{S^n}} (1-\|\mathbf{x}\|)^z d\mathbf{x}.$$

Now let us introduce several basic results about $\varpi(\overline{S^n})$ and $\varpi(S^n)$.

**Lemma 4.3 (Haagerup and Munkholm, 1981).** *For $n \geq 2$, we have*

$$\frac{n-1}{n^2} \leq \frac{\varpi(\overline{S^{n+1}})}{\varpi(\overline{S^n})} \leq \frac{1}{n}.$$

**Proof.** First of all it follows by (4.18) and (4.19) that

$$\varpi(\overline{S^n}) = \int_{\overline{S^n}} (1-\|\mathbf{x}\|^2)^{-\frac{n+1}{2}}\, d\mathbf{x} \tag{4.20}$$

and

$$n\,\varpi(\overline{S^{n+1}}) = \int_{\overline{S^n}} (1 - \|\mathbf{x}\|^2)^{-\frac{n}{2}}\, d\mathbf{x}. \tag{4.21}$$

Next we try to prove

$$\frac{n-1}{n}\,\varpi(\overline{S^n}) = \int_{\overline{S^n}} (1 - \|\mathbf{x}\|^2)^{-\frac{n-1}{2}}\, d\mathbf{x}. \tag{4.22}$$

Let us define a vector field

$$V(\mathbf{x}) = (1 - \|\mathbf{x}\|^2)^{-\frac{n-1}{2}}\,\mathbf{x}$$

for $\|\mathbf{x}\| < 1$. Then we have

$$\operatorname{div}(V(\mathbf{x})) = (1 - \|\mathbf{x}\|^2)^{-\frac{n-1}{2}} + (n-1)(1 - \|\mathbf{x}\|^2)^{-\frac{n+1}{2}}$$

and, when $\mathbf{x} \in \partial(\overline{S^n})$ and $\mathbf{n}$ is an outward normal of $\overline{S^n}$ at $\mathbf{x}$

$$\langle V(\mathbf{x}), \mathbf{n}\rangle = \tfrac{1}{n}\,(1 - \|\mathbf{x}\|^2)^{-\frac{n-1}{2}}.$$

Let $F$ denote one of the $n+1$ facets of $\overline{S^n}$. For $\mathbf{x} \in F$, we let $|\mathbf{x}|$ denote the Euclidean distance between $\mathbf{x}$ and the center of gravity of $F$, and write $r = (1 - n^{-2})^{1/2}$. By *Gauss' divergence formula*

$$\int_{\overline{S^n}} \operatorname{div}(V(\mathbf{x}))\, d\mathbf{x} = \int_{\partial(\overline{S^n})} \langle V(\mathbf{x}), \mathbf{n}\rangle\, ds$$

we get

$$\begin{aligned}
\varphi(n, \tfrac{n-1}{2}) + (n-1)\,\varphi(n, \tfrac{n+1}{2}) &= \frac{1}{n} \int_{\partial(\overline{S^n})} (1 - \|\mathbf{x}\|^2)^{-\frac{n-1}{2}}\, ds \\
&= \frac{n+1}{n} \int_F (r^2 - |\mathbf{x}|^2)^{-\frac{n-1}{2}}\, ds \\
&= \frac{n+1}{n} \int_{\overline{S^{n-1}}} (r^2 - r^2\|\mathbf{x}\|^2)^{-\frac{n-1}{2}}\, r^{n-1}\, d\mathbf{x} \\
&= \frac{n+1}{n} \int_{\overline{S^{n-1}}} (1 - \|\mathbf{x}\|^2)^{-\frac{n-1}{2}}\, d\mathbf{x} \\
&= \frac{n+1}{n}\, \varphi(n-1, \tfrac{n-1}{2}).
\end{aligned}$$

Applying (4.21) and (4.20) to this formula, we get

$$\begin{aligned}
\varphi(n, \tfrac{n-1}{2}) &= \frac{n+1}{n}\, \varphi(n-1, \tfrac{n-1}{2}) - (n-1)\,\varphi(n, \tfrac{n+1}{2}) \\
&= \frac{(n+1)(n-1)}{n}\, \varpi(\overline{S^n}) - (n-1)\,\varpi(\overline{S^n}) \\
&= \frac{n-1}{n}\, \varpi(\overline{S^n}),
\end{aligned}$$

which is (4.22). Finally, by (4.20), (4.21), and (4.22), we can deduce

$$\tfrac{n-1}{n}\,\varpi(\overline{S^n}) \le n\,\varpi(\overline{S^{n+1}}) \le \varpi(\overline{S^n}),$$

which proves the lemma. □

**Lemma 4.4 (Haagerup and Munkholm, 1981).** *Let $S$ be a simplex with all vertices on the boundary of $H_p^n$, let $r$ denote the Euclidean distance between its center of gravity and the origin, and let $f(x)$ be a concave function defined on* $(0, 1]$. *Then*

$$\frac{1}{v_n(S)}\int_S f(1-\|\mathbf{x}\|^2)\,d\mathbf{x} \le \frac{1}{v_n(\overline{S^n})}\int_{\overline{S^n}} f((1-r^2)(1-\|\mathbf{x}\|^2))\,d\mathbf{x}$$

*whenever both integrals converge.*

**Proof.** For convenience, let us abbreviate the left-hand side and the right-hand side of the inequality to $J_1$ and $J_2$, respectively. Let $\mathbf{v}_0, \mathbf{v}_1, \ldots, \mathbf{v}_n$ be the vertices of $S$ and write

$$T^n = \left\{(t_0, t_1, \ldots, t_n) \in R^{n+1}:\ t_i \ge 0,\ \sum t_i = 1\right\}.$$

Then we have

$$J_1 = \int_{T^n} f\left(1 - \left\|\sum t_i \mathbf{v}_i\right\|^2\right) d\mu,$$

where $\mu$ is the normalized Lebesgue measure on $T^n$. Since $\mu$ is invariant under the transformation $t_i \to t_{\pi(i)}$ for any permutation $\pi$ of $\{0, 1, \ldots, n\}$, we have

$$J_1 = \int_{T^n} f\left(1 - \left\|\sum t_{\pi(i)} \mathbf{v}_i\right\|^2\right) d\mu.$$

Let $E$ denote the formation of mean values over all such permutations $\pi$. Then, by the concavity assumption, we have

$$\begin{aligned} J_1 &= E\left(\int_{T^n} f\left(1 - \left\|\sum t_{\pi(i)} \mathbf{v}_i\right\|^2\right) d\mu\right) \\ &\le \int_{T^n} f\left(E\left(1 - \left\|\sum t_{\pi(i)} \mathbf{v}_i\right\|^2\right)\right) d\mu. \end{aligned} \tag{4.23}$$

Since

$$\left\|\sum t_{\pi(i)} \mathbf{v}_i\right\|^2 = \sum_{i\ne j} t_{\pi(i)} t_{\pi(j)} \langle \mathbf{v}_i, \mathbf{v}_j\rangle + \sum t_i^2,$$

$$\begin{aligned} E\left(t_{\pi(i)} t_{\pi(j)}\right) &= \frac{1}{n(n+1)} \sum_{k\ne l} t_k t_l \\ &= \frac{1}{n(n+1)}\left(1 - \sum t_i^2\right), \end{aligned}$$

and

$$\sum_{i \neq j} \langle \mathbf{v}_i, \mathbf{v}_j \rangle = \left\| \sum \mathbf{v}_i \right\|^2 - \sum \|\mathbf{v}_i\|^2$$
$$= (n+1)^2 r^2 - (n+1),$$

by (4.23) we get

$$J_1 \leq \int_{T^n} f\left(1 - \sum t_i^2 - \tfrac{(n+1)^2 r^2 - (n+1)}{n(n+1)} \left(1 - \sum t_i^2\right)\right) d\mu$$
$$= \int_{T^n} f\left(\tfrac{n+1}{n} (1 - r^2) \left(1 - \sum t_i^2\right)\right) d\mu. \tag{4.24}$$

On the other hand, we note that equality in (4.24) will hold if $S$ is regular. Therefore, applying (4.24) to $\overline{S^n}$ and to $g(\mathbf{x}) = f((1 - r^2)\mathbf{x})$, we get

$$J_2 = \int_{T^n} f\left(\tfrac{n+1}{n} (1 - r^2) \left(1 - \sum t_i^2\right)\right) d\mu. \tag{4.25}$$

The lemma follows from (4.24) and (4.25). □

**Lemma 4.5 (Haagerup and Munkholm, 1981).** *For every simplex $S$ in $H_p^n$, we have*

$$\varpi(S) \leq \varpi\left(\overline{S^n}\right).$$

**Proof (A sketch).** The two-dimensional case is trivial. For $n = 3$, we can deduce the assertion by *Lobachevsky's volume formula* (see Thurston, 1977).

Assume that the assertion is true for some $n \geq 3$, we proceed to show it for $n + 1$. Let $S = \text{conv}\{\mathbf{v}_0, \mathbf{v}_1, \ldots, \mathbf{v}_{n+1}\}$ be an $(n+1)$-dimensional simplex inscribed in $H_p^{n+1}$ with $\mathbf{v}_{n+1} = \mathbf{e}_{n+1}$ and write $S' = \text{conv}\{g(\mathbf{v}_0), g(\mathbf{v}_1), \ldots, g(\mathbf{v}_n)\}$ ($g(\mathbf{x})$ is defined just above (4.18)). Then we define

$$c_n = \frac{n\, \varpi(\overline{S^{n+1}})}{\varpi(\overline{S^n})}$$

and

$$f(t) = t^{-\frac{n}{2}} - c_n t^{-\frac{n+1}{2}}.$$

By Lemma 4.3, we get

$$c_n \geq \frac{n-1}{n} > \frac{n(n+2)}{(n+1)(n+3)}$$

for $n \geq 3$ and therefore $f(t)$ is strictly concave on $(0, 1]$. Then it follows by (4.18), (4.19), Lemma 4.4, and the inductive assumption that

$$\begin{aligned} n\,\varpi(S) - c_n\,\varpi(S') &= \int_{S'} (1 - \|\mathbf{x}'\|^2)^{-\frac{n}{2}}\, d\mathbf{x}' - c_n \int_{S'} (1 - \|\mathbf{x}'\|^2)^{-\frac{n+1}{2}}\, d\mathbf{x}' \\ &= \int_{S'} f(1 - \|\mathbf{x}'\|^2)\, d\mathbf{x}' \\ &\leq \int_{\overline{S^n}} f((1 - r^2)(1 - \|\mathbf{x}\|^2))\, d\mathbf{x} \\ &= (1 - r^2)^{-\frac{n}{2}}\, n\,\varpi(\overline{S^{n+1}}) - c_n (1 - r^2)^{-\frac{n+1}{2}}\, \varpi(\overline{S^n}) \\ &\leq (1 - r^2)^{-\frac{n}{2}} \left( n\,\varpi(\overline{S^{n+1}}) - c_n\,\varpi(\overline{S^n}) \right) \\ &= 0. \end{aligned}$$

Thus, we have

$$n\,\varpi(S) \leq c_n\,\varpi(\overline{S^n}) = n\,\varpi(\overline{S^{n+1}}),$$

which proves the lemma. □

**Remark 4.2.** This result was first conjectured by W. Thurston (1977).

**Lemma 4.6 (Haagerup and Munkholm, 1981).**

$$\varpi(\overline{S^n}) \leq \left( \frac{n+1}{n-1} \right)^{\frac{n+1}{2}} \frac{\sqrt{n}}{n!}.$$

**Proof (A sketch).** By a routine analytic argument (considering the second order derivative of $\log \varphi(n, z)$ and applying the Cauchy–Schwarz inequality), it can be shown that $\varphi(n, z)$, as a function of $z$, is *logconvex*[5] on $[0, \frac{n+1}{2}]$. Thus, we have

$$\frac{\varphi(n, \frac{n-1}{2})}{\varphi(n, 0)} \leq \left( \frac{\varphi(n, \frac{n+1}{2})}{\varphi(n, \frac{n-1}{2})} \right)^{\frac{n-1}{2}}.$$

Since $\varphi(n, 0) = v_n(\overline{S^n})$, by (4.20) and (4.22), we have

$$\frac{\varpi(\overline{S^n})}{v_n(\overline{S^n})} \leq \frac{n}{n-1} \left( \frac{\varphi(n, \frac{n+1}{2})}{\varphi(n, \frac{n-1}{2})} \right)^{\frac{n-1}{2}} = \left( \frac{n}{n-1} \right)^{\frac{n+1}{2}}.$$

---

[5] A function $f(x)$ is logconvex in $[a, b]$ if

$$f(\lambda x_1 + (1 - \lambda) x_2) \leq f(x_1)^{\lambda} f(x_2)^{1-\lambda}$$

holds whenever $x_1,\ x_2 \in [a, b]$ and $0 \leq \lambda \leq 1$.

On the other hand, it is known that

$$v_n(\overline{S^n}) = \left(\frac{n+1}{n}\right)^{\frac{n+1}{2}} \frac{\sqrt{n}}{n!}.$$

Therefore, we get

$$\varpi(\overline{S^n}) \leq \left(\frac{n+1}{n-1}\right)^{\frac{n+1}{2}} \frac{\sqrt{n}}{n!},$$

which proves the lemma. □

**Proof of Theorem 4.3.** Let $\widetilde{I^n}$ denote a largest cube inscribed in $H_p^n$ and assume that $\Im = \{S_1, S_2, \ldots, S_{\tau_n}\}$ is a triangulation of $\widetilde{I^n}$. Then we have

$$v_n(\widetilde{I^n}) = \left(\frac{2}{\sqrt{n}}\right)^n = \left(\frac{4}{n}\right)^{\frac{n}{2}}$$

and

$$\varpi(\widetilde{I^n}) = \sum_{i=1}^{\tau_n} \varpi(S_i). \tag{4.26}$$

Next we observe that

$$\begin{aligned}
\frac{1}{v_n(\widetilde{I^n})} \int_{\widetilde{I^n}} \|\mathbf{x}\|^2 d\mathbf{x} &= \frac{1}{v_n(\widetilde{I^n})} \int_{\widetilde{I^n}} \sum x_i^2 \, d\mathbf{x} \\
&= \frac{\frac{1}{3} \cdot 2 \cdot \left(\frac{1}{\sqrt{n}}\right)^3 \cdot n \cdot \frac{v_n(\widetilde{I^n})\sqrt{n}}{2}}{v_n(\widetilde{I^n})} \\
&= \tfrac{1}{3}.
\end{aligned}$$

Therefore, for any convex function $f(t)$, we can deduce

$$\frac{1}{v_n(\widetilde{I^n})} \int_{\widetilde{I^n}} f(\|\mathbf{x}\|^2) \, d\mathbf{x} \geq f\left(\frac{1}{v_n(\widetilde{I^n})} \int_{\widetilde{I^n}} \|\mathbf{x}\|^2 d\mathbf{x}\right) = f(\tfrac{1}{3}).$$

By choosing $f(t) = (1-t)^{-\frac{n+1}{2}}$, we get

$$\frac{\varpi(\widetilde{I^n})}{v_n(\widetilde{I^n})} \geq \left(1 - \frac{1}{3}\right)^{-\frac{n+1}{2}} = \left(\frac{3}{2}\right)^{\frac{n+1}{2}}, \tag{4.27}$$

and therefore

$$\varpi(\widetilde{I^n}) \geq \left(\frac{4}{n}\right)^{\frac{n}{2}} \left(\frac{3}{2}\right)^{\frac{n+1}{2}}. \tag{4.28}$$

Then, applying (4.28), Lemma 4.5, and Lemma 4.6 to (4.26), we get

$$\tau_n \geq \frac{\varpi(\widetilde{I^n})}{\varpi(\overline{S^n})} \geq \frac{\left(\frac{4}{n}\right)^{\frac{n}{2}}\left(\frac{3}{2}\right)^{\frac{n+1}{2}}}{\left(\frac{n+1}{n-1}\right)^{\frac{n+1}{2}}\frac{\sqrt{n}}{n!}}$$

$$\geq \frac{1}{2}\left(\frac{3}{2}\right)^{\frac{n}{2}}\eta_n.$$

The theorem is proved. □

**Remark 4.3.** By a detailed computation, Smith (2000) was able to improve (4.27) into

$$\lim_{n\to\infty}\left(\frac{\varpi(\widetilde{I^n})}{v_n(\widetilde{I^n})}\right)^{\frac{1}{n}} = 1.261522510\ldots.$$

## 4.4 Lower-dimensional cases

In 1976 Mara introduced a convenient way to enumerate the vertices of $\overline{I^n}$. If $\mathbf{a} = (a_1, a_2, \ldots, a_n)$ is a vertex of $\overline{I^n}$, then we denote it by a number

$$a = \sum_{i=1}^{n} 2^{i-1} a_i.$$

Clearly the numbers $0, 1, \ldots, 2^n - 1$ exactly represent the $2^n$ vertices of $\overline{I^n}$. Then we denote the simplex with vertices $\mathbf{a}, \mathbf{b}, \ldots, \mathbf{p}$ by $[a, b, \ldots, p]$. With this notation, he discovered a triangulation of $\overline{I^4}$ with 16 simplices

$$\begin{array}{lll} S_1 = [0,1,2,4,8], & S_2 = [1,2,3,7,11], & S_3 = [1,2,4,7,14], \\ S_4 = [1,2,4,8,14], & S_5 = [1,2,7,11,14], & S_6 = [1,2,8,11,14], \\ S_7 = [1,4,5,7,13], & S_8 = [1,4,7,13,14], & S_9 = [1,4,8,13,14], \\ S_{10} = [1,7,11,13,14], & S_{11} = [1,8,9,11,13], & S_{12} = [1,8,11,13,14], \\ S_{13} = [2,4,6,7,14], & S_{14} = [2,8,10,11,14], & S_{15} = [4,8,12,13,14], \end{array}$$

and

$$S_{16} = [7, 11, 13, 14, 15].$$

Therefore, we have

$$\tau_4 \leq 16. \tag{4.29}$$

On the other hand, we claim

$$\tau_4 \geq 16. \tag{4.30}$$

Let $\Im = \{S_1, S_2, \ldots, S_{\tau_4}\}$ be a triangulation for $I^4$, let $\{F_{i,1}, F_{i,2}\}$, $i = 1, 2, 3$, and 4, be the four pairs of opposite facets of $I^4$, and let $\Im_{i,j}$ denote the subset of $\Im$ which belong to $F_{i,j}$. By Theorem 3.7*, we can deduce

$$v_4(S_i) = \frac{m_i}{4!}, \tag{4.31}$$

where $m_i$ only takes three possible values 1, 2, and 3. Since $\Im_{i,j}$ induces a triangulation on $F_{i,j}$, by Theorem 4.1 we get

$$\text{card}\{\Im_{i,j}\} = l_{i,j}, \tag{4.32}$$

where $l_{i,j}$ only takes two possible values 5 and 6. In addition, when

$$\text{card}\{\Im_{i,j}\} = 5,$$

there is a simplex $S_k \in \Im$ satisfying the following conditions.

1 *It belongs to one and only one facet $F_{i,j}$ of $I^n$.*

2
$$v_4(S_k) = \tfrac{2}{4!} = \tfrac{1}{12}. \tag{4.33}$$

Now let us consider three cases.

**Case 1.** card$\{\Im_{i,j}\} = 5$ *holds for all indices i and j*. Let $S_1, S_2, \ldots, S_{10}$ be the ten different simplices belonging to either $F_{1,1}$ or $F_{1,2}$. In addition, for each facet $F_{i,j}$, $i \geq 2$, by condition 1 we get an extra simplex. Therefore, $\Im$ has at least six extra simplices and thus

$$\tau_4 \geq 10 + 6 = 16.$$

**Case 2.** card$\{\Im_{i,1}\}$ + card$\{\Im_{i,2}\} \leq 11$ *holds for* $i = 1, 2, 3,$ *and* 4, *and equality holds for* $i = 1$. Let $S_1, S_2, \ldots, S_{11}$ be the eleven simplices which belong to either $F_{1,1}$ or $F_{1,2}$. Then we have

$$\sum_{i=1}^{11} v_4(S_i) = \tfrac{1}{2}. \tag{4.34}$$

In addition, by conditions 1 and 2 we have three extra simplices $S_{12}$, $S_{13}$, and $S_{14}$ belonging to $\{F_{2,1}, F_{2,2}\}$, $\{F_{3,1}, F_{3,2}\}$, and $\{F_{4,1}, F_{4,2}\}$, respectively, and satisfying

$$\sum_{i=12}^{14} v_4(S_i) = \tfrac{1}{4}. \tag{4.35}$$

Then, by (4.34) and (4.35), we get

$$\sum_{i=15}^{\tau_4} v_4(S_i) = 1 - \tfrac{1}{2} - \tfrac{1}{4} = \tfrac{1}{4}$$

and therefore

$$\tau_4 \geq 14 + \tfrac{1}{4}/\tfrac{1}{8} = 16.$$

**Case 3.** $\mathrm{card}\{\Im_{1,1}\} + \mathrm{card}\{\Im_{1,2}\} = 12$. Let $S_1, S_2, \ldots, S_{12}$ be the 12 simplices which belong to either $F_{1,1}$ or $F_{1,2}$. Then we have

$$\sum_{i=1}^{12} v_4(S_i) = \tfrac{1}{2}$$

and therefore by (4.31)

$$\tau_4 \geq 12 + \tfrac{1}{2}/\tfrac{1}{8} = 16.$$

As a conclusion of these cases, we have proved (4.30) and therefore the following theorem.

**Theorem 4.4 (Mara, 1976; Cottle, 1982; Sallee, 1982a; and Lee, 1985).**

$$\tau_4 = 16.$$

**Remark 4.4.** Lee's proof was based on some deep results about the $f$-vectors and $h$-vectors. Our proof here is based on Sallee's arguments.

**Theorem 4.5 (Hughes, 1993; Hughes and Anderson, 1993 and 1996).**

$$\tau_5 = 67, \qquad \tau_6 = 308, \qquad \tau_7 = 1493.$$

So far these are the only known results about the exact values of $\tau_n$. As one can imagine, Theorem 4.5 was proved by complicated linear and integer programs, with computer aid. It is neither possible nor in our interest to discuss such long proofs in this book. We refer the interested readers to the original papers.

Besides triangulations, let us introduce another interesting concept. Let $\Re = \{S_1, S_2, \ldots, S_k\}$ be a set of simplices. If

$$I^n = \bigcup_{i=1}^{k} S_i$$

and

$$\mathrm{int}(S_i) \bigcap \mathrm{int}(S_j) = \emptyset$$

holds for all distinct $i$ and $j$, then we say that $\Re$ is a *decomposition* of $I^n$. Similar to $\tau_n$, we define

$$\varphi_n = \min_{\Re} \Big\{ \mathrm{card}\{\Re\} \Big\},$$

where the minimum is over all decompositions of $I^n$. It was proved by Mara (1976), Sallee (1982a), and Hughes and Anderson (1996) that

$$\varphi_3 = \tau_3 = 5,$$

$$\varphi_4 = \tau_4 = 16,$$

and

$$\varphi_5 = \tau_5 = 67,$$

respectively. In fact, the proof arguments for Theorem 4.1 and Theorem 4.4 can show the first two cases. So far these are the only known results about the exact values of $\varphi_n$. There are many open problems about $\tau_n$ and $\varphi_n$. Here we only mention one of them.

**Problem 4.1.** Does $\varphi_n = \tau_n$ hold for all $n$?

For some polytopes, even in three-dimensional space, the answer to the similar problem is negative (see Below, Brehm, De Loera, and Richter-Gebert, 2000). Now let us end this chapter with the following remark.

**Remark 4.5.** As it was pointed out by Mara (1976), to determine the minimal triangulation of $I^n$ is closely related with minimizing the number of pivot steps in simplicial algorithms for finding approximate fixed points. Besides its own interest, this is another motive for studying the cube triangulations.

# 5

# 0/1 polytopes

## 5.1 Introduction

A 0/1 *polytope* is a convex hull of a subset of the vertices of the unit cube

$$I^n = \left\{ \mathbf{x} \in E^n : \ 0 \leq x_i \leq 1 \right\}.$$

In the planar case, the only 0/1 polygons are a point, a segment, a triangle, and a square. However, in higher dimensions the situation turns out to be extraordinarily complicated. For example, according to Aichholzer (2000), there are 1 226 525 different classes of 0/1 polytopes in five dimensions, with respect to 0/1 *equivalence*. Another example, let $\xi(n, m)$ denote the average volume of the 0/1 polytopes in $E^n$ and with $m$ vertices. It was shown by Dyer, Füredi, and McDiarmid (1992) that, let $\epsilon$ be any positive number and write $\varsigma = 2/\sqrt{e}$

$$\lim_{n\to\infty} \xi(n, m) = \begin{cases} 1 & \text{if } m \geq (\varsigma+\epsilon)^n, \\ 0 & \text{if } m \leq (\varsigma-\epsilon)^n. \end{cases}$$

In fact, the higher-dimensional 0/1 polytopes are rich in structure. Besides their own geometric and combinatorial interest, they do provide intuitive models for coding theory, combinatorial optimization, etc.

There are several fundamental problems about the geometry and the combinatorics of 0/1 polytopes. For example:

**Problem 5.1.** Determine or estimate the number of the different classes (with respect to a fixed equivalence) of all $n$-dimensional 0/1 polytopes.

**Problem 5.2.** Determine or estimate the maximal number of the $k$-faces of an $n$-dimensional 0/1 polytope.

These natural problems are simple or even trivial when $n$ is small. However, in higher dimensions they are indeed challenging and fascinating. So far our knowledge about them is very limited.

Perhaps the next problem is not as natural as the previous ones. However, it is one of the key problems in coding theory.

**Problem 5.3.** Given integers $n$ and $s$. What is the maximal number $A(n, s)$ such that there is an $n$-dimensional $0/1$ polytope with $A(n, s)$ vertices and the minimal distance between them is not smaller than $\sqrt{s}$?

Therefore deep study of $0/1$ polytopes will provide better understanding of coding theory. In this chapter, we will introduce some basic results about $0/1$ polytopes.

## 5.2 0/1 polytopes and coding theory

Let $F_2$ denote the *binary field* and let $\Omega_n$ denote the set of the $2^n$ vertices of $I^n$; that is

$$\Omega_n = \Big\{(x_1, x_2, \ldots, x_n) : \; x_i \in F_2\Big\}.$$

It is easy to see that, for any subset $X$ of $\Omega_n$

$$V(\mathrm{conv}\{X\}) = X.$$

Thus, every subset $X$ of $\Omega_n$ can be the vertices of a $0/1$ polytope, possibly not $n$-dimensional. In addition, for any two points $\mathbf{x}$ and $\mathbf{y}$ of $\Omega_n$

$$\|\mathbf{x}, \mathbf{y}\| = \sqrt{\mathrm{card}\{i : \; x_i \neq y_i\}}.$$

On the other hand, if we treat $\Omega_n$ as a linear space over $F_2$ and define a *Hamming metric*

$$\|\mathbf{x}, \mathbf{y}\|_H = \mathrm{card}\Big\{i : \; x_i \neq y_i\Big\}$$

on it, then we get an $n$-dimensional *binary Hamming space* $H_2^n$. Clearly, we have

$$\|\mathbf{x}, \mathbf{y}\| = \sqrt{\|\mathbf{x}, \mathbf{y}\|_H}, \tag{5.1}$$

whenever both $\mathbf{x}$ and $\mathbf{y}$ belong to $\Omega_n$.

Usually, a point $\mathbf{c} \in H_2^n$ is called a *binary codeword*, a subset $C$ of $H_2^n$ is called a *binary code*, and the minimum *Hamming distance* between distinct points in $C$ is called the *separation* of $C$, denoted by $s(C)$. In addition, the number of the nonzero coordinates of a codeword is called its *weight.*

For convenience, a code of length $n$, size $m$, and separation $s$ is called an *$(n, m, s)$-code*. Then we can restate Problem 5.3 as follows.

**Problem 5.3*.** Given $n$ and $s$. What is the maximal number $A(n, s)$ such that there is a code $C$ in $H_2^n$ with cardinality $A(n, s)$ and separation $s$?

Roughly speaking, an information transmission process can be described as follows. First, design a code $C$ and encode the information into codewords. Second, transmit the codewords through a channel to a receiver. Since the channel may add errors, the received words (in $H_2^n$) perhaps are not the sent ones. Third, design a decoder to eliminate the errors. In this step, a received word $\mathbf{w}$ will be replaced by a codeword $\mathbf{c} \in C$ satisfying

$$\|\mathbf{w}, \mathbf{c}\|_H = \min\left\{\|\mathbf{w}, \mathbf{w}'\|_H : \mathbf{w}' \in C\right\}.$$

It is easy to imagine that, if $s = s(C)$ is relatively large, then the errors caused by the transmission can be eliminated more easily. On the other hand, if card$\{C\}$ is relatively large, then the code is more efficient. To measure the efficiency of a code, we define

$$\text{ş}(C) = \frac{\log_2(\text{card}\{C\})}{n}$$

and call it the *information rate* of the code $C$. Based on the above arguments, it is easy to see that Problem 5.3* is indeed a key problem in coding theory.

Let us start with some basic results about $A(n, s)$. First of all, it is obvious that

$$A(n, 1) = 2^n$$

and

$$A(n, n) = 2.$$

Second, if $C$ is a binary $(n, m, s)$-code with $m = A(n, s)$ and for $i = 0$ and 1 define

$$C_i = \left\{\mathbf{c} \in C : \ c_1 = i\right\},$$

then $C_0$ will reduce to an $(n-1, m_0, s)$-code with a suitable $m_0$ and $C_1$ will reduce to an $(n-1, m_1, s)$-code with a suitable $m_1$. Since one of them has a cardinality not smaller than $A(n, s)/2$, we get

$$A(n, s) \leq 2\, A(n-1, s).$$

Third, if $C$ is a binary $(n, m, 2k-1)$-code with $m = A(n, 2k-1)$, by adding an *overall parity check* to each codeword, we can produce an $(n+1, m, 2k)$-code. On the other hand, suppose that $C$ is a binary $(n+1, m, 2k)$-code with

$m = A(n+1, 2k)$, by puncturing $C$ in a position at which two codewords disagree, we get an $(n, m, 2k-1)$-code with $m = A(n+1, 2k)$. Thus we have

$$A(n, 2k-1) = A(n+1, 2k). \tag{5.2}$$

Now we introduce one lower bound and three upper bounds for $A(n, s)$.

**Theorem 5.1 (The Gilbert–Varshamov bound).**

$$A(n, s) \geq \frac{2^n}{\sum_{i=0}^{s-1} \binom{n}{i}}.$$

**Proof.** Let $C$ be an $(n, m, s)$-code with $m = A(n, s)$. For every codeword $\mathbf{c} \in C$, we define

$$B_{\mathbf{c}} = \left\{ \mathbf{w} \in H_2^n : \ \|\mathbf{c}, \mathbf{w}\|_H < s \right\}$$

and

$$B_{\mathbf{c},i} = \left\{ \mathbf{w} \in H_2^n : \ \|\mathbf{c}, \mathbf{w}\|_H = i \right\}.$$

By the maximum assumption on $m$, it follows that

$$H_2^n = \bigcup_{\mathbf{c} \in C} B_{\mathbf{c}}. \tag{5.3}$$

In addition, it is easy to see that

$$\text{card}\{B_{\mathbf{c},i}\} = \binom{n}{i}$$

and

$$\text{card}\{B_{\mathbf{c}}\} = \text{card}\left\{ \bigcup_{i=0}^{s-1} B_{\mathbf{c},i} \right\} = \sum_{i=0}^{s-1} \binom{n}{i}. \tag{5.4}$$

Therefore, by (5.3) and (5.4), we get

$$A(n, s) \geq \frac{\text{card}\{H_2^n\}}{\text{card}\{B_{\mathbf{c}}\}} = \frac{2^n}{\sum_{i=0}^{s-1} \binom{n}{i}}.$$

Theorem 5.1 is proved. □

As a counterpart of this lower bound, we have the following upper bound.

**Theorem 5.2 (The Hamming bound).**

$$A(n, s) \leq \frac{2^n}{\sum_{i=0}^{s'-1} \binom{n}{i}},$$

*where* $s' = [(s-1)/2]$.

**Proof.** Let $C$ be an $(n, m, s)$-code with $m = A(n, s)$. Similar to the previous proof, for every codeword $\mathbf{c} \in C$ we define

$$B'_{\mathbf{c}} = \left\{ \mathbf{w} \in H_2^n : \ \|\mathbf{c}, \mathbf{w}\|_H < s' \right\}.$$

Since $s(C) = s$ and $s' = [(s-1)/2]$, it is easy to see that

$$B'_{\mathbf{c}} \bigcap B'_{\mathbf{d}} = \emptyset \tag{5.5}$$

holds for every pair of distinct codewords $\mathbf{c}$ and $\mathbf{d}$ of $C$. In addition, similar to (5.4), we can deduce

$$\mathrm{card}\{B'_{\mathbf{c}}\} = \sum_{i=0}^{s'-1} \binom{n}{i}.$$

Therefore, by (5.5) we obtain

$$A(n, s) \leq \frac{\mathrm{card}\{H_2^n\}}{\mathrm{card}\{B'_{\mathbf{c}}\}} = \frac{2^n}{\sum_{i=0}^{s'-1} \binom{n}{i}}.$$

Theorem 5.2 is proved. □

The Hamming bound is also known as the *sphere packing bound*. As a counterpart, Theorem 5.1 is also known as the *sphere covering bound*. The best-known result about $A(n, s)$, obtained by ideas of packing and covering, is the following theorem.

**Theorem 5.3 (The Elias bound).** *Assume that $r$ is an integer satisfying both $r \leq n/2$ and $r^2 - nr + ns/2 > 0$. Then*

$$A(n, s) \leq \frac{ns}{2r^2 - 2nr + ns} \cdot \frac{2^n}{\sum_{i=0}^{r} \binom{n}{i}}.$$

To prove this theorem we need two technical lemmas.

**Lemma 5.1.** *For each pair of subsets $C_1$ and $C_2$ of $H_2^n$ there is a $\mathbf{w} \in H_2^n$ such that*

$$\frac{\mathrm{card}\{(\mathbf{w} + C_1) \cap C_2\}}{\mathrm{card}\{C_1\}} \geq \frac{\mathrm{card}\{C_2\}}{2^n}.$$

**Proof.** Assume that $\mathrm{card}\{(\mathbf{x} + C_1) \bigcap C_2\}$ as a function of $\mathbf{x}$ attends its maximum at $\mathbf{x} = \mathbf{w}$. Then we have

$$\begin{aligned}
\mathrm{card}\left\{(\mathbf{w} + C_1) \bigcap C_2\right\} &\geq \frac{1}{2^n} \sum_{\mathbf{x} \in H_2^n} \mathrm{card}\left\{(\mathbf{x} + C_1) \bigcap C_2\right\} \\
&= \frac{1}{2^n} \sum_{\mathbf{x} \in H_2^n} \sum_{\mathbf{c}_1 \in C_1} \sum_{\mathbf{c}_2 \in C_2} \mathrm{card}\{\{\mathbf{x} + \mathbf{c}_1\} \cap \{\mathbf{c}_2\}\}
\end{aligned}$$

$$= \frac{1}{2^n} \sum_{\mathbf{c}_1 \in C_1} \sum_{\mathbf{c}_2 \in C_2} \sum_{\mathbf{x} \in H_2^n} \mathrm{card}\{\{\mathbf{x} + \mathbf{c}_1\} \cap \{\mathbf{c}_2\}\}$$

$$= \frac{1}{2^n} \sum_{\mathbf{c}_1 \in C_1} \sum_{\mathbf{c}_2 \in C_2} 1$$

$$= \frac{1}{2^n} \mathrm{card}\{C_1\} \cdot \mathrm{card}\{C_2\}.$$

The lemma is proved. □

**Lemma 5.2.** *If the weights of the words of an* $(n, m', s)$*-code* $C$ *have an upper bound* $r$ *with* $r \leq n/2$*, then*

$$m' \leq \frac{ns}{2r^2 - 2nr + ns}.$$

**Proof.** Let us list the codewords of the code as rows of an $m' \times n$ matrix and, for $j = 0$ and 1, let $p_{ij}$ denote the number of occurrences of the symbol $j$ in the $i$th column of the matrix. Then we have

$$p_{i0} + p_{i1} = m'$$

and, by the weight assumption

$$\sum_{i=1}^{n} p_{i0} \geq m'(n - r). \tag{5.6}$$

For convenience, we denote the left-hand side of (5.6) by $p$. Therefore, we get

$$\sum_{i=1}^{n} p_{i0}^2 \geq \frac{1}{n} \left( \sum_{i=1}^{n} p_{i0} \right)^2 = \frac{p^2}{n}.$$

Based on these preparations it can be deduced that

$$\sum_{\mathbf{c}, \mathbf{c}' \in C} \|\mathbf{c}, \mathbf{c}'\|_H = \sum_{i=1}^{n} \sum_{j=0}^{1} p_{ij}(m' - p_{ij})$$

$$= nm'^2 - \sum_{i=1}^{n} \left( p_{i0}^2 + p_{i1}^2 \right)$$

$$= nm'^2 - \sum_{i=1}^{n} \left( 2p_{i0}^2 + m'^2 - 2m' p_{i0} \right)$$

$$\leq nm'^2 - \left( \frac{2p^2}{n} + nm'^2 - 2m' p \right)$$

$$= 2m' p - \frac{2p^2}{n}.$$

In addition, by (5.6) and the assumption $r \le n/2$, we get

$$p \ge m'(n-r) \ge \tfrac{1}{2}nm'$$

and thus, by comparing the values of a quadratic form

$$\sum_{\mathbf{c},\mathbf{c}' \in C} \|\mathbf{c},\mathbf{c}'\|_H \le 2m'p - \frac{2p^2}{n} \le 2m'^2 r\left(1-\frac{r}{n}\right). \tag{5.7}$$

On the other hand, we have

$$\sum_{\mathbf{c},\mathbf{c}' \in C} \|\mathbf{c},\mathbf{c}'\|_H \ge m'(m'-1)s. \tag{5.8}$$

It follows by (5.7) and (5.8) that

$$m'(m'-1)s \le 2m'^2 r\left(1-\frac{r}{n}\right)$$

and therefore

$$m' \le \frac{ns}{2r^2-2nr+ns}.$$

The lemma is proved. □

**Proof of Theorem 5.3.** Let $C_2$ be a binary $(n, m, s)$-code with $m = A(n, s)$ and define

$$C_1 = \left\{\mathbf{x} \in H_2^n : \|\mathbf{o},\mathbf{x}\|_H \le r\right\}.$$

By Lemma 5.1, there is a suitable $(n, m', s)$-code

$$C_3 = (C_1 + \mathbf{w}) \bigcap C_2$$

satisfying

$$m' = \mathrm{card}\{C_3\} \ge \frac{m\sum_{i=0}^{r}\binom{n}{i}}{2^n}.$$

Then applying Lemma 5.2 to $C_3$, we get

$$\frac{m\sum_{i=0}^{r}\binom{n}{i}}{2^n} \le m' \le \frac{ns}{2r^2-2nr+ns}$$

and thus

$$m \le \frac{ns}{2r^2-2nr+ns} \cdot \frac{2^n}{\sum_{i=0}^{r}\binom{n}{i}}.$$

The theorem is proved. □

As has been mentioned before, to determine or to estimate the values of $A(n, s)$ is a key problem in coding theory. Besides the upper bounds introduced above, there are still several others, such as *the Plotkin bound, the Griesmer bound, the Johnson bound, the linear programming bound,* etc. (see van Lint, 1982 and Pless, Huffman, and Brualdi, 1998). The linear programming bound, though it is hard to get an exact expression, perhaps is the most interesting one. Let $K_j(x)$ denote the *Krawtchouk polynomial* of degree $j$; that is

$$K_j(x) = \sum_{i=0}^{j} (-1)^i \binom{x}{i}\binom{n-x}{j-i},$$

where

$$\binom{x}{i} = \frac{x(x-1)\cdots(x-i+1)}{i!}.$$

Then the linear programming bound can be stated as follows.

**Theorem 5.4 (Delsarte, 1972 and, 1973).** *If s is even*

$$A(n,s) \le \max\left\{ \sum_{l=0}^{n} a_l :\ a_0 = 1;\ a_l = 0 \text{ for } 1 \le l \le s \text{ or } l \text{ is odd}; \right.$$
$$\left. a_l \ge 0;\ \sum_{l=0}^{n} a_l K_j(l) \ge 0 \text{ for } 0 \le j \le n \right\}.$$

Clearly, we can apply (5.2) if $s$ is odd. When $n$ is relatively small, this method is powerful. However, when $n$ is large, it is difficult for application. For technical reasons, we will not prove it here. To end this section we list some known values of $A(n,s)$ in the following table. For more about $A(n,s)$ we refer to Pless, Huffman, and Brualdi (1998), Sloane (1977) or van Lint (1982).

| $n$ | 5 | 6 | 7 | 8 | 9 | 10 | 11 | 12 | 13 | 14 | 15 |
|---|---|---|---|---|---|---|---|---|---|---|---|
| $A(n,3)$ | 4 | 8 | 16 | 20 | 40 | 72 | 144 | 256 | 512 | 1024 | 2048 |
| $A(n,5)$ | 2 | 2 | 2 | 4 | 6 | 12 | 24 | 32 | 64 | 128 | 256 |
| $A(n,7)$ | – | – | 2 | 2 | 2 | 2 | 4 | 4 | 8 | 16 | 32 |

## 5.3 Classification

Let $\phi(n)$ denote the number of the $n$-dimensional 0/1 polytopes reduced from $\overline{I^n}$. It is easy to see that

$$\phi(n) < \sum_{i=n+1}^{2^n} \binom{2^n}{i} < 2^{2^n}. \tag{5.9}$$

On the other hand – note that any $(n-1)$-dimensional subset of a facet and a nonempty set of the opposite facet will produce an $n$-dimensional set – we can deduce

$$\phi(n) \geq \phi(n-1) \cdot \sum_{i=1}^{2^{n-1}} \binom{2^{n-1}}{i} = \left(2^{2^{n-1}} - 1\right) \cdot \phi(n-1) \tag{5.10}$$

and therefore

$$\phi(n) \geq \prod_{i=1}^{n} \left(2^{2^{i-1}} - 1\right) = \prod_{i=0}^{n-1} \left(1 - 2^{-2^i}\right) \prod_{i=0}^{n-1} 2^{2^i} \sim c \cdot 2^{2^n}, \tag{5.11}$$

where

$$c = \frac{1}{2} \prod_{i=0}^{\infty} \left(1 - 2^{-2^i}\right).$$

Comparing (5.11) with (5.9), we note that these bounds are reasonably good, although they are possibly not optimal.

There are several types of classification for 0/1 polytopes based on different equivalent relations; for example, the classifications based on *combinatorial equivalence*, *affine equivalence*, *congruence*, or 0/1 *equivalence*. Needless to say, the affine equivalence and the congruence are two fundamental concepts in Geometry and the combinatorial equivalence is basic to understanding polytopes. Only the 0/1 equivalence is a relatively new concept restricted to 0/1 polytopes. In this section, we will introduce some known results about these classifications.

Let $\mathcal{F}_P$ denote the *face lattice* of a polytope $P$; that is, the set of all faces of $P$ partially ordered by inclusion. Two poytopes $P_1$ and $P_2$ are combinatorially equivalent if $\mathcal{F}_{P_1}$ is *isomorphic* to $\mathcal{F}_{P_2}$. The combinatorial equivalence has two basic properties.

1 *If $P_1$ is equivalent with $P_2$ and $P_2$ is equivalent with $P_3$, then $P_1$ is equivalent with $P_3$.*
2 *If $\varphi$ is an isomorphism from $\mathcal{F}_{P_1}$ to $\mathcal{F}_{P_2}$, then*

$$\dim\{\varphi(F)\} = \dim\{F\}$$

*holds for every face $F$ of $P_1$ and therefore*

$$\mathrm{card}\Big\{F \in \mathcal{F}_{P_1} : \ \dim\{F\} = k\Big\} = \mathrm{card}\Big\{F \in \mathcal{F}_{P_2} : \ \dim\{F\} = k\Big\}$$

*holds for every $k$ satisfying* $0 \le k \le n$.

In the two-dimensional case, there are only two classes of $0/1$ polygons with respect to the combinatorial equivalence, one is represented by a triangle and the other is represented by a square. In three dimensions, the situation is much more complicated. By applying the second property, we can see from Figure 5.1 that there are exactly eight different classes of three-dimensional $0/1$ polytopes with respect to the combinatorial equivalence.

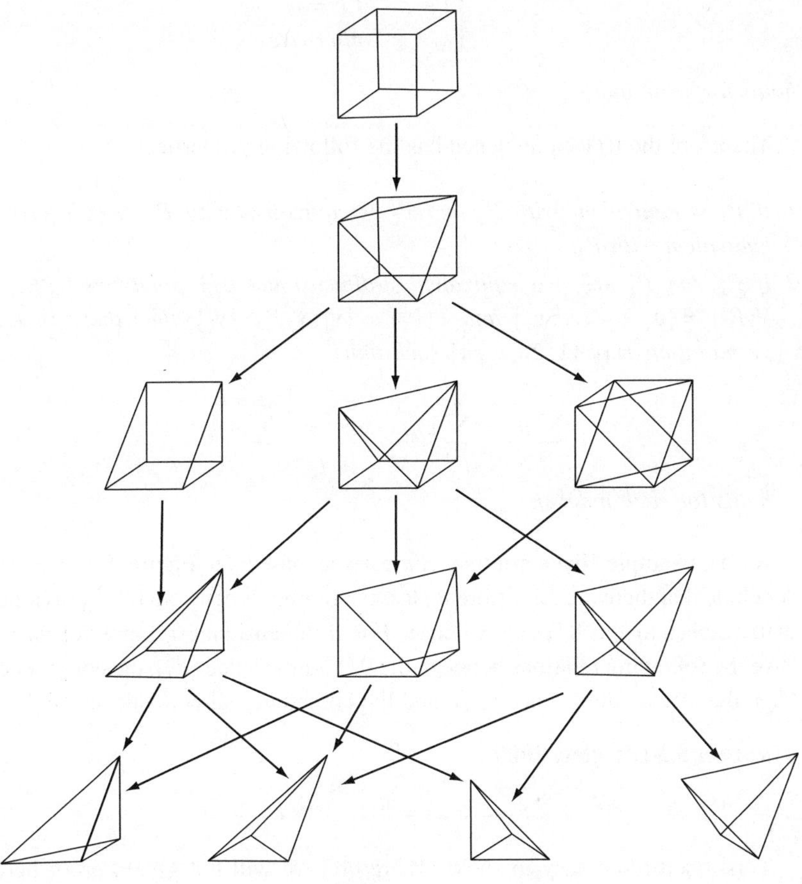

Figure 5.1

Two $0/1$ polytopes $P_1$ and $P_2$ are $0/1$ equivalent if one can be transformed into the other by a symmetry of the unit cube $\overline{I^n}$. It is known that any symmetry of $\overline{I^n}$ can be represented as a product of transformations of the following two types.

**Type I.** $\mathbf{x} \longrightarrow \mathbf{y}$, *where*

$$y_k = \begin{cases} x_i & \text{if } k = j, \\ x_j & \text{if } k = i, \\ x_k & \text{otherwise} \end{cases}$$

*holds for some index pair* $\{i, j\}$.

**Type II.** $\mathbf{x} \longrightarrow \mathbf{y}$, *where*

$$y_k = \begin{cases} 1 - x_i & \text{if } k = i, \\ x_k & \text{otherwise} \end{cases}$$

*holds for some index i.*

Therefore the $0/1$ equivalence has the following properties.

1 *If $P_1$ is equivalent with $P_2$ and $P_2$ is equivalent with $P_3$, then $P_1$ is equivalent with $P_3$.*
2 *If $P_1$ and $P_2$ are two equivalent n-dimensional $0/1$ polytopes with $V(P_1) = \{\mathbf{u}_1, \mathbf{u}_2, \ldots, \mathbf{u}_k\}$ and $V(P_2) = \{\mathbf{v}_1, \mathbf{v}_2, \ldots, \mathbf{v}_k\}$, then there is a permutation $\pi$ of $\{1, 2, \ldots, n\}$ such that*

$$\sum_{i=1}^{k} u_{ij} = \sum_{i=1}^{k} v_{i\pi(j)} \quad or \quad k - \sum_{i=1}^{k} v_{i\pi(j)}$$

*holds for each index j.*

As an example, by a routine comparison based on Figure 5.1, we can conclude that there are 12 different classes of three-dimensional $0/1$ polytopes with respect to the $0/1$ equivalence. For $n$-dimensional $0/1$ polytopes, we have the following relations between the $0/1$ equivalence ($E_1$), the congruence ($E_2$), the affine equivalence ($E_3$), and the combinatorial equivalence ($E_4$).

**Theorem 5.5 (Ziegler, 2000).**

$$E_1 \Longrightarrow E_2 \Longrightarrow E_3 \Longrightarrow E_4.$$

This assertion is easy to show. Therefore, we will not give a proof here. However, the converse to any of the three implications is false. Clearly, the first two tetrahedra of the last row in Figure 5.1 are affine equivalent but not congruent. To show the other cases, we have the following examples.

**Example 5.1.** Let $S_1$ be a five-dimensional simplex with vertices $\mathbf{u}_1 = (0,0,0,0,0)$, $\mathbf{u}_2 = (0,0,1,1,0)$, $\mathbf{u}_3 = (0,1,0,1,0)$, $\mathbf{u}_4 = (1,0,0,1,0)$, $\mathbf{u}_5 = (0,1,1,0,0)$, and $\mathbf{u}_6 = (0,1,1,0,1)$, and let $S_2$ be a five-dimensional simplex with vertices $\mathbf{v}_1 = (0,0,0,0,0)$, $\mathbf{v}_2 = (0,0,1,1,0)$, $\mathbf{v}_3 = (0,1,0,1,0)$, $\mathbf{v}_4 = (0,1,1,0,0)$, $\mathbf{v}_5 = (1,0,0,1,0)$, and $\mathbf{v}_6 = (1,0,0,1,1)$. It is routine to verify that

$$\|\mathbf{u}_i - \mathbf{u}_j\| = \|\mathbf{v}_i - \mathbf{v}_j\|$$

holds for all index pairs $\{i, j\}$. Thus $S_1$ and $S_2$ are congruent. However, by the second property of the 0/1 equivalence, we can easily deduce that $S_1$ and $S_2$ are not 0/1 equivalent.

**Example 5.2.** Let $P_1$ be a five-dimensional polytope with vertices $\mathbf{u}_1 = (0,0,0,0,0)$, $\mathbf{u}_2 = (1,0,0,0,0)$, $\mathbf{u}_3 = (0,1,0,0,0)$, $\mathbf{u}_4 = (0,0,1,0,0)$, $\mathbf{u}_5 = (0,0,0,1,0)$, $\mathbf{u}_6 = (0,0,0,0,1)$, and $\mathbf{u}_7 = (1,1,1,1,1)$, and let $P_2$ be a five-dimensional polytope with vertices $\mathbf{v}_1 = (0,0,0,0,0)$, $\mathbf{v}_2 = (1,1,0,0,0)$, $\mathbf{v}_3 = (0,1,1,0,0)$, $\mathbf{v}_4 = (0,0,1,1,0)$, $\mathbf{v}_5 = (0,0,0,1,1)$, $\mathbf{v}_6 = (1,0,0,0,1)$, and $\mathbf{v}_7 = (1,1,1,1,1)$. In fact, both $P_1$ and $P_2$ are *bipyramids* over four-dimensional simplices. Therefore, they are combinatorially equivalent. However, since the main diagonals of $P_1$ and $P_2$ are divided by the simplices in the ratios $1:4$ and $2:3$, respectively, they are not affinely equivalent.

Let $\phi_1(n)$, $\phi_2(n)$, $\phi_3(n)$, and $\phi_4(n)$ denote the numbers of the different classes of the $n$-dimensional 0/1 polytopes with respect to the 0/1 equivalence, the congruence, the affine equivalence, and the combinatorial equivalence, respectively. It follows by Theorem 5.5 that

$$\phi_4(n) \le \phi_3(n) \le \phi_2(n) \le \phi_1(n) \le \phi(n). \tag{5.12}$$

For large $n$ to determine the exact values of $\phi_i(n)$ or even $\phi(n)$ is a very hard job. So far, our knowledge of this kind is very limited. We list the known ones in the following table.

| $n$ | $\phi(n)$ | $\phi_1(n)$ | $\phi_2(n)$ | $\phi_3(n)$ | $\phi_4(n)$ |
|---|---|---|---|---|---|
| 2 | 5 | 2 | 2 | 2 | 2 |
| 3 | 151 | 12 | 12 | 8 | 8 |
| 4 | 60879 | 347 | 347 | ?? | 172 |
| 5 | 4292660729 | 1226525 | ?? | ?? | ?? |

It follows by (5.9) and (5.12) that

$$\phi_i(n) < 2^{2^n} \tag{5.13}$$

holds for all $i = 1, 2, 3$, and 4. This upper bound is certainly not optimal, since, on the one hand, many of the 0/1 polytopes are lower-dimensional; on the other hand, many of the full-dimensional ones are equivalent. However, so far no essentially better upper bound for $\phi_i(n)$ is known. As a counterpart of (5.13), we have the following lower bound for $\phi_i(n)$.

**Theorem 5.5 (Ziegler, 2000).** *When $n \geq 6$, we have*

$$\phi_i(n) \geq 2^{2^{n-2}}$$

*for all $i = 1, 2, 3$, and 4.*

**Proof.** By (5.12), to prove the theorem it is sufficient to show

$$\phi_4(n) \geq 2^{2^{n-2}}. \tag{5.14}$$

Let $F_i^0$ and $F_i^1$ denote the facets of $\overline{I^n}$ given by $x_i = 0$ and $x_i = 1$, respectively. For convenience, we will call $F_n^0$ the *bottom facet*, call $F_n^1$ the *top facet*, and call all the others *vertical facets* of $\overline{I^n}$. Let $\mathcal{P}_n$ denote the family of the 0/1 polytopes $P$ reduced from $\overline{I^n}$ and satisfying the following conditions.

1 *It contains the whole bottom facet of $\overline{I^n}$.*
2 *It contains both $\mathbf{e}_n = (0, 0, \ldots, 1)$ and $\mathbf{e} = (1, 1, \ldots, 1)$.*
3 *It contains neither $\mathbf{e}_n + \mathbf{e}_1 = (1, 0, \ldots, 1)$ nor $\mathbf{e} - \mathbf{e}_1 = (0, 1, \ldots, 1)$.*

Clearly, all the polytopes contained in $\mathcal{P}_n$ are $n$-dimensional and

$$\text{card}\{\mathcal{P}_n\} = \sum_{i=0}^{2^{n-1}-4} \binom{2^{n-1}-4}{i} = 2^{2^{n-1}-4}. \tag{5.15}$$

Assume that $\mathcal{P}_n$ can be divided into combinatorially equivalent classes $\mathcal{C}_1$, $\mathcal{C}_2, \ldots, \mathcal{C}_k$, we claim that

$$\text{card}\{\mathcal{C}_j\} \leq 2^{n-1} \cdot (n-1)! \tag{5.16}$$

holds for all $j = 1, 2, \ldots, k$.

To prove (5.16) let us start with looking at the facial structure of an individual polytope $P \in \mathcal{P}_n$. By condition 2, it follows that each vertical facet of $\overline{I^n}$ induces a facet for $P$ which is adjacent to the cubic facet $F_n^0$. Thus, its vertices $\mathbf{v} \notin F_n^0$ are completely determined by its vertical facets. On the other hand, $P$ has no other cubic facet except $F_n^0$. To see this, if $H$ is a hyperplane which contains at least $2^{n-1}$ vertices of $I^n$, then by symmetry $-H$ will contain at least $2^{n-1}$ vertices as well. Thus $H$ contains either a whole facet of $I^n$ or the origin. By conditions 1, 2, and 3, it follows that $P$ has no other cubic facet.

Assume that $P_1$ and $P_2$ are two polytopes of $\mathcal{P}_n$ and they are combinatorially equivalent. If $\varphi$ is an isomorphism from $\mathcal{F}_{P_1}$ to $\mathcal{F}_{P_2}$, then it follows from the above observations that $\varphi$ induces a symmetry of $\overline{I^{n-1}}$, which is determined by $\varphi(\mathbf{u}) \longrightarrow \mathbf{v}$, where $\mathbf{u}, \mathbf{v} \in V(F_n^0)$. Since the order of the symmetry group of $\overline{I^{n-1}}$ is $2^{n-1} \cdot (n-1)!$, then (5.16) is proved.

By (5.15) and (5.16), it follows that

$$\phi_4(n) \geq \frac{2^{2^{n-1}-4}}{2^{n-1} \cdot (n-1)!} = \frac{2^{2^{n-2}} \cdot 2^{2^{n-2}}}{2^{n+3} \cdot (n-1)!}. \tag{5.17}$$

Writing $f(n) = 2^{2^{n-2}}$ and $g(n) = 2^{n+3} \cdot (n-1)!$, it is easy to see that $f(6) > g(6)$, $f(n) = f(n-1)^2$

$$g(n) = 2(n-1) \cdot g(n-1) \leq g(n-1)^2,$$

and therefore

$$f(n) \geq g(n)$$

provided $n \geq 6$. Thus, when $n \geq 6$, by (5.17), we get

$$\phi_4(n) \geq 2^{2^{n-2}}.$$

The theorem is proved. □

## 5.4 The number of facets

In this section, we will discuss some known results about Problem 5.2. Let $f(n, k)$ denote the maximal number of the $k$-dimensional faces of an $n$-dimensional 0/1 polytope, and especially abbreviate $f(n, n-1)$ to $f(n)$. The known exact values of $f(n)$ can be listed as follows (see Ziegler, 2000).

| $n$ | 2 | 3 | 4 | 5 |
|---|---|---|---|---|
| $f(n)$ | 4 | 8 | 16 | 40 |

It is interesting to notice that $f(5) > 2^5$. Thus we may imagine that, when $n$ is large, to determine or to estimate $f(n)$ is a very hard job.

Let $\{\mathbf{e}_1, \mathbf{e}_2, \ldots, \mathbf{e}_n\}$ be a standard basis of $E^n$, and write $\mathbf{e} = \sum_{i=1}^{n} \mathbf{e}_i$. It is easy to see that

$$T_n = \text{conv}\{\mathbf{e}_1,\ \mathbf{e}-\mathbf{e}_1, \ldots, \mathbf{e}_n,\ \mathbf{e}-\mathbf{e}_n\}$$

is centrally symmetric with respect to the center of $\overline{I^n}$ and therefore it is an $n$-dimensional 0/1 cross polytope. By this example, we get

$$f(n) \geq 2^n. \tag{5.18}$$

Based on this simple observation, we can deduce the following result.

**Lemma 5.3 (Kortenkamp, Richter–Gebert, Sarangarajan, and Ziegler, 1997).** *For $i = 1$ and 2, if $\overline{I^{n_i}}$ has an $n_i$-dimensional centrally symmetric 0/1 polytope with $f_i$ facets, then $\overline{I^{n_1+n_2}}$ has an $(n_1+n_2)$-dimensional centrally symmetric 0/1 polytope with $f_1 f_2$ facets.*

Writing $n = n_1 + n_2$, it is routine to check that

$$Q_1 = \left\{\mathbf{x} \in \overline{I^n}:\ x_{n_1} = x_{n_1+1} = \cdots = x_n\right\}$$

is affinely equivalent with $\overline{I^{n_1}}$

$$Q_2 = \left\{\mathbf{x} \in \overline{I^n}:\ x_1 = \cdots = x_{n_1} = 1 - x_{n_1+1}\right\}$$

is affinely equivalent with $\overline{I^{n_2}}$, and $Q_1$ intersects $Q_2$ at the center of $\overline{I^n}$. If $P_i$ is an $n_i$-dimensional centrally symmetric polytope with $f_i$ facets and satisfying $V(P_i) \subseteq V(Q_i)$, for $i = 1$ and 2, then it can be shown that $\text{conv}\{P_1 \cap P_2\}$ is an $n$-dimensional centrally symmetric 0/1 polytope with $f_1 f_2$ facets.

Let $f^*(n)$ denote the maximal number of the facets of an $n$-dimensional centrally symmetric 0/1 polytope. Then, by Lemma 5.3, we get

$$f(n) \geq f^*(n) \geq f^*(n_1) \cdot f^*(n_2). \tag{5.19}$$

On the other hand, Christof and Reinelt (2001) discovered a 13-dimensional centrally symmetric 0/1 polytope with 17 464 356 facets and therefore

$$f^*(13) \geq 17464356 > 3.6^{13}. \tag{5.20}$$

Thus, by (5.19) and (5.20), Ziegler (2000) got the following lower bound for $f(n)$.

**Theorem 5.6.** *When $n$ is sufficiently large, we have*

$$f(n) \geq f^*(n) \geq 3.6^n.$$

**Remark 5.1.** Based on the number of the facets of certain ten-dimensional centrally symmetric 0/1 polytope, Kortenkamp, Richter-Gebert, Sarangarajan, and Ziegler (1997) were able to deduce

$$f(n) \geq f^*(n) \geq 2.76^n$$

for sufficiently large $n$, the first lower bound better than (5.18).

The following lower bound was proved by a random method. Since its proof is too complicated to be introduced here, we only cite the statement.

**Theorem 5.7 (Bárány and Pór, 2001; Gatzouras, Giannopoulos, and Markoulakis, 2005).** *There is an absolute constant c such that*

$$f(n) \geq \left(\frac{c\,n}{\log^2 n}\right)^{n/2}.$$

Now, let us introduce an upper bound for $f(n)$.

**Theorem 5.8 (Fleiner, Kaibel, and Rote, 2000).** *There is a positive number c such that*

$$f(n) \leq c \cdot (n-2)!.$$

**Proof.** First of all, let us make some observations.

1 *The volume of any n-dimensional* $0/1$ *polytope P is an integer multiple of* $1/n!$. *In particular,* $v(P) \geq 1/n!$.
2 *Let P be an n-dimensional polytope with facets* $F_1, F_2, \ldots, F_p$ *and let* $\Gamma$ *denote a projection from* $E^n$ *to an* $(n-1)$*-dimensional hyperplane H. Then*

$$\sum_{i=1}^{p} v_{n-1}(\Gamma(F_i)) = 2 \cdot v_{n-1}(\Gamma(P)).$$

3 *Let* $\Gamma_i$ *denote the projection from* $E^n$ *to* $H_i = \{\mathbf{x} : x_i = 0\}$. *If a hyperplane H contains a subset Q with* $0 < v_{n-1}(Q) < \infty$, *then it has a normal vector* $\mathbf{n}$ *of the form*

$$\mathbf{n} = \Big(\pm v_{n-1}(\Gamma_1(Q)),\ \pm v_{n-1}(\Gamma_2(Q)), \ldots, \pm v_{n-1}(\Gamma_n(Q))\Big).$$

Assume that $P$ is an $n$-dimensional $0/1$ polytope with facets $F_1, F_2, \ldots, F_p$ and abbreviate $v_{n-1}(\Gamma_i(F_j))$ to $\mu_{ij}$. Then it follows by Observations 1 and 3 that $F_j$ has an integral normal vector of the form

$$\mathbf{n}_j = (n-1)! \cdot \Big(\pm \mu_{1j},\ \pm \mu_{2j}, \ldots, \pm \mu_{nj}\Big).$$

Let $\|\cdot\|^\star$ denote the $\ell_1$ norm; that is

$$\|\mathbf{x}\|^\star = \sum_{i=1}^{n} |x_i|.$$

Then, by Observation 2

$$\begin{aligned}\sum_{j=1}^{p} \|\mathbf{n}_j\|^\star &= (n-1)! \sum_{j=1}^{p}\sum_{i=1}^{n} \mu_{ij} \\ &= (n-1)! \sum_{i=1}^{n}\sum_{j=1}^{p} \mu_{ij} \\ &\le (n-1)! \sum_{i=1}^{n} 2 \\ &= 2\cdot n!. \end{aligned} \tag{5.21}$$

If $p \le (n-2)!$, then there is nothing to prove. If $p > (n-2)!$, then we proceed to show

$$p \le c\cdot (n-2)!. \tag{5.22}$$

For convenience, we write

$$G_r = \left\{\mathbf{z} \in Z^n : \sum |z_i| \le r\right\},$$

$$J_r = G_r \setminus G_{r-1},$$

and

$$\Sigma_r = \sum_{\mathbf{x}\in G_r} \|\mathbf{x}\|^\star = \sum_{i=0}^{r} i \cdot \mathrm{card}\{J_i\}.$$

It is easy to see that

$$\mathrm{card}\{G_r\} = v(G_r + I^n) \le \frac{2^n(r+\frac{n}{2})^n}{n!} = \frac{(2r+n)^n}{n!} \tag{5.23}$$

and

$$\mathrm{card}\{J_{r+1}\} \ge \mathrm{card}\{J_r\}.$$

Thus we have

$$\begin{aligned}\Sigma_r &= \frac{1}{2}\sum_{i=0}^{r}\Big(i\cdot \mathrm{card}\{J_i\} + (r-i)\cdot \mathrm{card}\{J_{r-i}\}\Big) \\ &\ge \frac{1}{2}\sum_{i=0}^{r}\frac{r}{2}\Big(\mathrm{card}\{J_i\} + \mathrm{card}\{J_{r-i}\}\Big) \\ &= \frac{r}{2}\sum_{i=1}^{r} \mathrm{card}\{J_r\} \\ &\ge \frac{r}{2}\,\mathrm{card}\{G_r\}. \end{aligned} \tag{5.24}$$

Let $k$ be a number satisfying

$$\operatorname{card}\{G_k\} \le p < \operatorname{card}\{G_{k+1}\}. \tag{5.25}$$

By the assumption $p > (n-2)!$, (5.23), and (5.25), we get

$$(n-2)! < \frac{(2k+n+2)^n}{n!}.$$

Therefore, for sufficiently large $n$, by Stirling's formula we obtain

$$\begin{aligned} k &> \frac{1}{2}\sqrt[n]{n!\cdot(n-2)!} - \frac{n}{2} - 1 \\ &> \frac{n}{8}\left(\frac{n-2}{4}\right)^{\frac{n-2}{n}} - \frac{n}{2} - 1 \\ &> d\cdot n^2, \end{aligned} \tag{5.26}$$

where $d$ is a suitable constant satisfying $0 < d < 1$.

Then, by (5.21), (5.25), (5.24), and (5.26), we get

$$\begin{aligned} 2\cdot n! &\ge \sum_{j=1}^{p} \|\mathbf{n}_j\|^\star \ge \Sigma_k + k\left(p - \operatorname{card}\{G_k\}\right) \\ &\ge \frac{k}{2}\operatorname{card}\{G_k\} + k\left(p - \operatorname{card}\{G_k\}\right) \\ &\ge \frac{k}{2}\left(\operatorname{card}\{G_k\} + p - \operatorname{card}\{G_k\}\right) \\ &\ge \frac{d\cdot n^2}{2}\cdot p \end{aligned}$$

and therefore

$$p \le \frac{4}{d}\cdot(n-2)!.$$

Thus we have proved (5.22) and the theorem. □

**Remark 5.2.** Before Theorem 5.8, based on Observation 1, Bárány (see Ziegler, 2000) proved

$$f(n) \le n! + 2n.$$

In fact, Fleiner, Kaibel, and Rote (2000) also obtained an upper bound for $f(n,k)$; that is

$$f(n,k) \le c\cdot(2(k+1))^{\frac{n(n-1)}{n+1}}\cdot(n-2)!.$$

As we can see, when $k = 0$ this bound is much worse than the trivial bound $2^n$, and when $k = n - 1$ it is not as good as Theorem 5.8.

**Remark 5.3.** As a conclusion of Theorem 5.7 and Theorem 5.8, we get

$$\left(\frac{c_1 n}{\log^2 n}\right)^{n/2} \leq f(n) \leq c_2 \cdot (n-2)!,$$

where $c_1$ and $c_2$ are constants.

# 6

# Minkowski's conjecture

## 6.1 Minkowski's conjecture

Let $C$ be an $n$-dimensional centrally convex body and let $X$ be a discrete set in $E^n$. If

$$E^n = \bigcup_{\mathbf{x} \in X} \left( C + \mathbf{x} \right)$$

and

$$\left( \operatorname{int}(C) + \mathbf{x} \right) \cap \left( \operatorname{int}(C) + \mathbf{y} \right) = \emptyset$$

holds for all distinct points $\mathbf{x}$ and $\mathbf{y}$ of $X$, then we call $C + X$ a *tiling* of $E^n$ and call $C$ a *tile*. For example, both a regular hexagon and a square are two-dimensional tiles. In fact, up to linear transformations, they are the only tiles in $E^2$. This fact was discovered by Fedorov in 1885. In 1908, it was proved by Voronoi (1908/1909) that there are exact five different types of three-dimensional tiles, the parallelotope (also known as *parallelopiped*), the *hexagonal prism*, the *rhombic dodecahedron*, the *elongated dodecahedron*, and the *truncated octahedron* (see Erdös, Gruber, and Hammer, 1989). According to Delone (1929) and Štogrin (1975), there are 52 different types of tiles in four-dimensional Euclidean space. In higher dimensions, as we can imagine, the situation becomes extremely complicated and our knowledge is very limited. Nevertheless, for every $n \geq 2$, it is obvious that $I^n$ is a tile in $E^n$.

Let $\mathbf{a}_1, \mathbf{a}_2, \ldots, \mathbf{a}_n$ be $n$ linearly independent vectors in $E^n$, then the set

$$\Lambda = \left\{ \sum_{i=1}^{n} z_i \mathbf{a}_i : \; z_i \in Z \right\}$$

is called an $n$-dimensional lattice and the set $\{\mathbf{a}_1, \mathbf{a}_2, \ldots, \mathbf{a}_n\}$ is called a basis for the lattice. For example, the set of all points of integer coordinates is a lattice. Clearly lattices are very regular discrete sets in $E^n$, periodic and

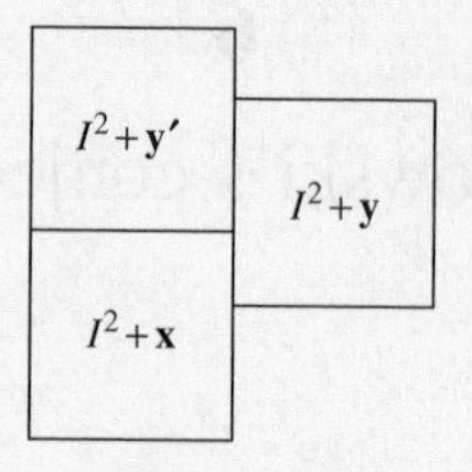


Figure 6.1

centrally symmetric. We will call $K+\Lambda$ a *lattice tiling* of $E^n$ if it is a tiling and if $\Lambda$ is a lattice[6]. There are thousands of references about tiling and lattice. We refer the interested readers to Engel (1993), and Erdös, Gruber, and Hammer (1989), and Schulte (1993).

Now let us observe a simple phenomenon. Assume that $X$ is a discrete set and $I^2+X$ is a tiling of the two-dimensional Euclidean plane $E^2$. If $I^2+\mathbf{x}$ touches $I^2+\mathbf{y}$ at its boundary, where $\mathbf{x}$ and $\mathbf{y}$ are two distinct points of $X$, then their intersection will be a whole edge or a part of an edge of $I^2+\mathbf{x}$. In the second case, since $I^2+X$ is a tiling of $E^2$, as we can see from Figure 6.1, there is another square $I^2+\mathbf{y}'$, which touches $I^2+\mathbf{x}$ at a whole edge. Thus, we get the following conclusion.

*If $I^2+X$ is a tiling of $E^2$, then there are two squares sharing a whole edge. Especially, if $I^2+\Lambda$ is a lattice tiling, then we have either $\{(z,0): z\in Z\}\subset\Lambda$ or $\{(0,z): z\in Z\}\subset\Lambda$.*

The three-dimensional case is much more complicated than the two-dimensional one, but is still elementary in nature. Let $I^3+X$ be a tiling of $E^3$ and, without loss of generality, assume that $\mathbf{o}\in X$. If $I^3+\mathbf{x}$ touches $I^3$ at its vertex $\mathbf{v}$, then $\mathbf{v}$ is either a relative interior point of a face of $I^3+\mathbf{x}$, or is a relative interior point of an edge of $I^3+\mathbf{x}$, or a vertex of $I^3+\mathbf{x}$. In the first case, by projecting all the cubes (except $I^3+\mathbf{x}$), which contain $\mathbf{v}$ to the hyperplane that contains the mentioned face of $I^3+\mathbf{x}$ and repeating the argument for the two-dimensional case, it can be shown that one of them touches $I^3$ at a whole face. If none of the cubes touching $I^3$ at $\mathbf{v}=(\frac{1}{2},\frac{1}{2},\frac{1}{2})$ shares a whole face with it, by a routine argument we can deduce that, up to permutations of coordinates, the tiling has three cubes $I^3+(1,0,\alpha_1)$, $I^3+(0,\alpha_2,1)$, and $I^3+(\alpha_3,1,0)$, where $0<\alpha_i<1$ holds for $i=1$, 2, and 3. Then the vertex $(\frac{1}{2},\alpha_2-\frac{1}{2},\frac{1}{2})$ of $I^3+(0,\alpha_2,1)$ will be a relative interior point of a face of

[6] It was proved by Venkov (see Zong, 1996) that there is a lattice tiling $K+\Lambda$ whenever $K$ is a tile.

$I^3 + (1, 0, \alpha_1)$ and therefore one of the cubes touches $I^3 + (0, \alpha_2, 1)$ at a whole face. Thus, we get the following conclusion.

*If $I^3 + X$ is a tiling of $E^3$, then two of the unit cubes share a whole face. Especially, if $I^3 + \Lambda$ is a lattice tiling, then we have either $\{(z, 0, 0) : z \in Z\} \subset \Lambda$, or $\{(0, z, 0) : z \in Z\} \subset \Lambda$, or $\{(0, 0, z) : z \in Z\} \subset \Lambda$.*

For convenience, we call two $n$-dimensional unit cubes a *twin* if they share a whole facet. Based on the above observations, we can make several conjectures, which will be the subjects of the remaining chapters of this book. In this chapter, we will deal with the following one.

**Minkowski's conjecture.** *Every lattice tiling $I^n + \Lambda$ of $E^n$ has a twin.*

## 6.2 An algebraic version

Clearly a lattice is an abelian group under addition. This is why we can get an algebraic version for Minkowski's conjecture.

As usual, let $G$ denote a group. For $\mathbf{g} \in G$, let $\langle \mathbf{g} \rangle$ denote the *cyclic group* generated by $\mathbf{g}$ and let $|\mathbf{g}|$ denote its *order*. Let us start with a basic result about the structure of the abelian groups.

**Lemma 6.1.** *A finitely generated abelian group $G$ is the direct sum of a finite number of cyclic groups.*

**Proof.** Let $\mathbf{0}$ denote the unit of $G$. Let $n$ denote the minimal cardinality of the generators and define

$$m = \min \left\{ |\mathbf{g}_1| \right\},$$

where the minimum is over all the possible generators $\{\mathbf{g}_1, \mathbf{g}_2, \dots, \mathbf{g}_n\}$ of $G$.

Based on induction, we assume that

$$G = \langle \mathbf{g}_1 \rangle + G^*,$$

where $\mathbf{g}_1$ satisfies $|\mathbf{g}_1| = m$ and $G^*$ is a direct sum of $n-1$ cyclic groups, say $\langle \mathbf{g}_2 \rangle, \dots, \langle \mathbf{g}_n \rangle$. If, on the contrary, $G$ is not a direct sum of $\langle \mathbf{g}_1 \rangle$ and $G^*$, then

$$z_1\mathbf{g}_1 + z_2\mathbf{g}_2 + \cdots + z_n\mathbf{g}_n = \mathbf{0}$$

holds for some suitable integers $z_i$ with

$$0 < z_1 < |\mathbf{g}_1|. \tag{6.1}$$

If $(z_1, z_2, \dots, z_n) = k$ (the *common divisor* of $z_1, z_2, \dots, z_n$) and $z_i = kz_i'$, then

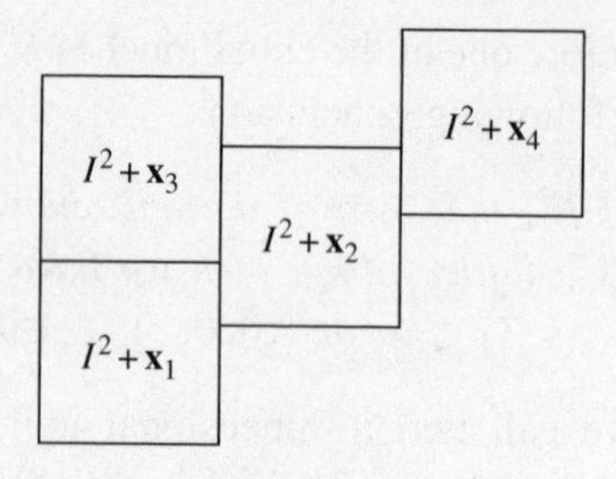


Figure 6.2

we have

$$(z'_1, z'_2, \ldots, z'_n) = 1.$$

It is well known in linear algebra that then we can construct a set of new generators $\{\mathbf{g}'_1, \mathbf{g}'_2, \ldots, \mathbf{g}'_n\}$ for $G$, where

$$\mathbf{g}'_1 = z'_1\mathbf{g}_1 + z'_2\mathbf{g}_2 + \cdots + z'_n\mathbf{g}_n.$$

Since

$$k\mathbf{g}'_1 = z_1\mathbf{g}_1 + z_2\mathbf{g}_2 + \cdots + z_n\mathbf{g}_n = \mathbf{0},$$

by the assumption on $|\mathbf{g}_1|$, we get

$$|\mathbf{g}_1| \leq k.$$

On the other hand, by (6.1) and the definition of $k$, we have

$$k \leq z_1 < |\mathbf{g}_1|.$$

By this contradiction, the lemma is proved. □

Let $I^n + \Lambda$ be a lattice tiling of $E^n$. Two cubes $I^n + \mathbf{u}$ and $I^n + \mathbf{v}$ are called *adjacent* if $|u_1 - v_1| = 1$ and $|u_i - v_i| < 1$ for all indices $i \neq 1$. In other words, two cubes are adjacent if their intersection is an $(n-1)$-dimensional set, which is perpendicular to the first coordinate axis. In fact, the adjacency between cubes does induce an equivalent relation on them, which will be one of the key ideas to deduce the algebraic version of Minkowski's conjecture.

Let $I^n + \mathbf{x}$ and $I^n + \mathbf{x}'$ be two cubes of the tiling. We say that they are *equivalent* to each other if there is a sequence of cubes $I^n + \mathbf{x}_1, I^n + \mathbf{x}_2, \ldots, I^n + \mathbf{x}_m$, where $\mathbf{x}_1 = \mathbf{x}$ and $\mathbf{x}_m = \mathbf{x}'$, such that $I^n + \mathbf{x}_i$ is adjacent to $I^n + \mathbf{x}_{i+1}$ for each $i = 1, 2, \ldots, m-1$. As shown in Figure 6.2, both $I^2 + \mathbf{x}_3$ and $I^2 + \mathbf{x}_4$ are equivalent with $I^2 + \mathbf{x}_1$. Clearly, the equivalence is well defined and similar relations can be defined with respect to the other axes. For convenience, we denote this equivalent relation by $\sim$.

By this equivalence, the unit cubes in $I^n + \Lambda$ are divided into different classes. It is easy to see that an individual cube cannot be shifted in the direction of the first axis, but the union of a whole class can be shifted in this direction. In other words, the union of the unit cubes which belong to the same class form a cylinder of infinite length. For convenience, we say a lattice is *rational* if all the coordinates of its points are rational. Otherwise, we say it is *irrational*. Based on this preparation, we can now prove the following lemma.

**Lemma 6.2 (Schmidt, 1933).** *If there is a lattice tiling $I^n + \Lambda$ without a twin, then we can find a rational lattice tiling $I^n + \Lambda'$ without a twin.*

**Proof.** Assume that $I^n + \Lambda$ is a lattice tiling of $E^n$, which has no twin, and $\Lambda$ is irrational. For convenience, we assume that the first coordinate of some lattice point is irrational and define

$$\Lambda^o = \left\{ \mathbf{x} \in \Lambda : \ \mathbf{x} \sim \mathbf{o} \right\}$$

and

$$\Lambda^\star = \left\{ \mathbf{x} \in \Lambda : \ x_1 \text{ is rational} \right\}.$$

Clearly, $\Lambda^\star$ is a sublattice of $\Lambda$ and $\Lambda^o \subseteq \Lambda^\star$. Since $\Lambda$ has a basis, the *quotient group* $\Lambda/\Lambda^\star$ is finitely generated. It follows by Lemma 6.1 that $\Lambda/\Lambda^\star$ has a basis. In addition, since $\alpha$ will be rational if $l\alpha$ is rational for some integer $l$

$$|\overline{\mathbf{u}}| = \infty$$

holds for all $\mathbf{u} \notin \Lambda^\star$. Thus $\Lambda$ can be written as a direct sum of two sublattices

$$\Lambda = \Gamma \oplus \Lambda^\star.$$

Let $\{\mathbf{u}_1, \mathbf{u}_2, \ldots, \mathbf{u}_r\}$ be a basis for $\Gamma$ and let $\{\mathbf{u}_{r+1}, \mathbf{u}_{r+2}, \ldots, \mathbf{u}_n\}$ be a basis for $\Lambda^\star$. Then $\{\mathbf{u}_1, \mathbf{u}_2, \ldots, \mathbf{u}_n\}$ will be a basis for $\Lambda$. In addition, the first coordinate $u_{i1}$ of $\mathbf{u}_i$ is irrational whenever $i \le r$ and is rational whenever $i \ge r+1$. Now we define

$$\mathbf{v}_i = \begin{cases} \mathbf{u}_i + \epsilon_i \mathbf{e}_1 & \text{if } i \le r, \\ \mathbf{u}_i & \text{otherwise,} \end{cases} \tag{6.2}$$

where $\epsilon_i$ are real numbers to be chosen in a moment, and define

$$\Lambda' = \left\{ \sum_{i=1}^{n} z_i \mathbf{v}_i : \ z_i \in Z \right\}.$$

Clearly, we have

$$\sum_{i=1}^{n} z_i \mathbf{v}_i = \sum_{i=1}^{n} z_i \mathbf{u}_i + \left( \sum_{i=1}^{r} z_i \epsilon_i \right) \mathbf{e}_1.$$

Thus, changing from $I^n + \Lambda$ to $I^n + \Lambda'$, the relative position of the cubes, which belong to the same class $\overline{\mathbf{u}} + \Lambda^\star$, does not change, but the whole class shifts $\sum_{i=1}^{r} z_i \epsilon_i$ in the direction of $\mathbf{e}_1$. This means that $I^n + \Lambda'$ is also a lattice tiling of $E^n$.

Let $\Omega$ be a tiling lattice for $I^n$ and let $\lambda(\Omega)$ denote the length of the shortest vectors of $\Omega \setminus \{\mathbf{o}\}$. Since the interior of $2I^n$ contains no other point of $\Omega$ but the origin $\mathbf{o}$, we get

$$\lambda(\Omega) \geq 1, \tag{6.3}$$

where equality holds if and only if $I^n + \Omega$ has a twin.

It follows by a routine argument that, for a fixed lattice $\Omega$ with a basis $\{\mathbf{a}_1, \mathbf{a}_2, \ldots, \mathbf{a}_n\}$ and for any fixed number $\delta > 0$, there is a corresponding number $\epsilon > 0$ such that

$$|\lambda(\Omega') - \lambda(\Omega)| < \delta \tag{6.4}$$

holds whenever $\Omega'$ is a lattice with a basis $\{\mathbf{b}_1, \mathbf{b}_2, \ldots, \mathbf{b}_n\}$ satisfying

$$\|\mathbf{a}_i - \mathbf{b}_i\| < \epsilon$$

for all indices $i$.

Now we take $\Omega = \Lambda$, $\mathbf{a}_i = \mathbf{u}_i$, and $\Omega' = \Lambda'$. Since $I^n + \Lambda$ has no twin, we get

$$\lambda(\Lambda) = 1 + 2\delta \tag{6.5}$$

for some $\delta > 0$. Then we have a corresponding number $\epsilon$ such that (6.4) holds. Now we choose

$$\epsilon_i = -u_{i1} + \gamma_i$$

for (6.2), where $\gamma_i$ are rational numbers satisfying

$$|\epsilon_i| = |\gamma_i - u_{i1}| < \epsilon.$$

Then the first coordinate of any point of $\Lambda'$ is rational and it follows by (6.4) and (6.5) that

$$\begin{aligned} \lambda(\Lambda') &> \lambda(\Lambda) - |\lambda(\Lambda) - \lambda(\Lambda')| \\ &\geq 1 + \delta > 1. \end{aligned}$$

Therefore, by (6.3), $I^n + \Lambda'$ is a lattice tiling of $E^n$ which has no twin.

The lemma can be proved by repeating this process at most $n$ times, each time dealing with one coordinate. □

Let $\Lambda$ be a rational lattice with a basis

$$\mathbf{a}_1 = \left(\frac{q_{11}}{p_{11}}, \frac{q_{12}}{p_{12}}, \ldots, \frac{q_{1n}}{p_{1n}}\right),$$
$$\mathbf{a}_2 = \left(\frac{q_{21}}{p_{21}}, \frac{q_{22}}{p_{22}}, \ldots, \frac{q_{2n}}{p_{2n}}\right),$$
$$\cdots$$
$$\mathbf{a}_n = \left(\frac{q_{n1}}{p_{n1}}, \frac{q_{n2}}{p_{n2}}, \ldots, \frac{q_{nn}}{p_{nn}}\right),$$

where $(q_{ij}, p_{ij}) = 1$ holds for all indices $i$ and $j$. Then we define[7]

$$d_i = \left[p_{1i}, p_{2i}, \ldots, p_{ni}\right],$$

$$\mathbf{v}_i = \tfrac{1}{d_i}\mathbf{e}_i,$$

$$P = \left\{\mathbf{x} \in E^n : 0 \le x_i \le \tfrac{1}{d_i}\right\},$$

and

$$\Gamma = \left\{\sum_{i=1}^{n} z_i \mathbf{v}_i : z_i \in Z\right\}.$$

It is easy to see that $P + \Gamma$ is a lattice tiling of $E^n$ and $\Lambda$ is a sublattice of $\Gamma$. Consequently, we get a quotient group

$$G = \Gamma/\Lambda = A_1 + A_2 + \cdots + A_n,$$

where

$$A_i = \left\{\overline{\mathbf{o}}, \overline{\mathbf{v}}_i, 2\overline{\mathbf{v}}_i, \ldots, (d_i - 1)\overline{\mathbf{v}}_i\right\}$$

and $\overline{\mathbf{v}}_i$ is the quotient of $\mathbf{v}_i$.

Now we proceed to deduce a connection between the group $G$ and the two systems $\overline{I^n} + \Lambda$ and $P + \Gamma$. If $\mathbf{u} \in \Lambda$, $\mathbf{v} \in \Gamma$, and

$$\left(\mathrm{int}(P) + \mathbf{v}\right) \cap \left(\overline{I^n} + \mathbf{u}\right) \neq \emptyset,$$

then we have

$$P + \mathbf{v} \subseteq \overline{I^n} + \mathbf{u}.$$

---

[7] As usual, here $[a_1, a_2, \ldots, a_n]$ denotes the *least common multiple* of $a_1, a_2, \ldots, a_n$.

Therefore, if one parallelotope $P+\mathbf{v}$ belongs to two of the unit cubes, then the point $\mathbf{v}$ can be represented in two distinct ways

$$\mathbf{v} = \mathbf{u} + \sum_{i=1}^{n} z_i \mathbf{v}_i, \tag{6.6}$$

where $\mathbf{u} \in \Lambda$ and $0 \le z_i \le d_i - 1$. On the other hand, if there is a parallelotope $P+\mathbf{v}$ does not belong to any of the unit cubes, then the point $\mathbf{v}$ cannot be represented in the above form. As a conclusion, $\overline{I^n}+\Lambda$ is a tiling of $E^n$ if and only if any point $\mathbf{v} \in \Gamma$ can be uniquely represented in the form of (6.6), therefore

$$G = A_1 \oplus A_2 \oplus \cdots \oplus A_n.$$

On the other hand, if $\overline{I^n}+\Lambda$ has a twin, say $\overline{I^n}$ and $\overline{I^n}+\mathbf{e}_i$, then we have

$$d_i \mathbf{v}_i = \mathbf{e}_i \in \Lambda$$

and thus $A_i$ is a cyclic group. Thus, by Lemma 6.2, we get an algebraic version for Minkowski's conjecture. For the convenience of its proof, we state it in multiplication form.

**Minkowski's conjecture (Hajós, 1941).** *Let $G$ be a finite abelian group with unit $\mathbf{1}$. If $\mathbf{g}_1, \mathbf{g}_2, \ldots, \mathbf{g}_n$ are elements of $G$ and $d_1, d_2, \ldots, d_n$ are positive integers such that each element $\mathbf{g}$ of $G$ can be uniquely written in the form*

$$\mathbf{g} = \prod_{i=1}^{n} \mathbf{g}_i^{z_i}, \quad 0 \le z_i \le d_i - 1,$$

*then $\mathbf{g}_i^{d_i} = \mathbf{1}$ holds for some $i$ with $1 \le i \le n$.*

## 6.3 Hajós' proof

Let $G$ be an abelian group under multiplication and let $\mathbf{1}$ denote its *unit*. For convenience, we will call a set $[\mathbf{g}]_m = \{\mathbf{1}, \mathbf{g}, \ldots, \mathbf{g}^{m-1}\}$, $\mathbf{g} \in G$, a *cyclic set*. Clearly, a cyclic subgroup is a special cyclic set.

**Lemma 6.3 (Hajós, 1941).** *Any cyclic subset $[\mathbf{g}]_m$ of $G$ may be decomposed into a direct product of cyclic subsets of prime orders. Moreover, $[\mathbf{g}]_m$ is a subgroup of $G$ if and only if one of the cyclic subsets is a subgroup.*

**Proof.** Assume that $m = p_1 p_2 \cdots p_k$, where $p_i$ are primes. It is known that every integer $l$, $0 \le l < m$, can be uniquely written as

$$l = c_0 + c_1 p_1 + c_2 p_1 p_2 + \cdots + c_{k-1} p_1 p_2 \cdots p_{k-1},$$

where $c_i$ are integers satisfying $0 \le c_i \le p_{i+1} - 1$. Therefore, $[\mathbf{g}]_m$ can be decomposed as a direct product

$$[\mathbf{g}]_m = [\mathbf{g}]_{p_1} \otimes [\mathbf{g}^{p_1}]_{p_2} \otimes \cdots \otimes [\mathbf{g}^{p_1 p_2 \cdots p_{k-1}}]_{p_k}.$$

If $[\mathbf{g}]_m$ is a subgroup of $G$, then we have

$$\mathbf{g}^m = \mathbf{g}^{p_1 p_2 \cdots p_k} = (\mathbf{g}^{p_1 p_2 \cdots p_{k-1}})^{p_k} = \mathbf{1}.$$

Therefore, $[\mathbf{g}^{p_1 p_2 \cdots p_{k-1}}]_{p_k}$ is a subgroup.

Now we proceed to show that, if $[\mathbf{g}]_m = \langle \mathbf{b} \rangle \otimes H$ holds for some cyclic group $\langle \mathbf{b} \rangle$ and a set $H$, then $[\mathbf{g}]_m$ is a cyclic group as well. Since $\mathbf{1} \in [\mathbf{g}]_m$, we have

$$\mathbf{1} = \mathbf{b}^{\gamma} \cdot \mathbf{h}$$

for some integer $\gamma$ and some element $\mathbf{h} \in H$. Then we get

$$\mathbf{b} = \mathbf{b}^{\gamma+1} \cdot \mathbf{h} \in [\mathbf{g}]_m$$

and therefore

$$\mathbf{b} = \mathbf{g}^s \tag{6.7}$$

for some positive integer $s \le m-1$. On the other hand, it follows by $[\mathbf{g}]_m = \langle \mathbf{b} \rangle \otimes H$ that $\mathbf{b} \cdot \mathbf{c} \in [\mathbf{g}]_m$ if $\mathbf{c} \in [\mathbf{g}]_m$. Therefore, if $l = rs + t$ and $0 \le t \le s-1$, by (6.7) we have

$$\mathbf{g}^l = \mathbf{g}^{rs} \cdot \mathbf{g}^t = \mathbf{b}^r \cdot \mathbf{g}^t \in [\mathbf{g}]_m$$

and finally

$$[\mathbf{g}]_m = \langle \mathbf{g} \rangle.$$

The lemma is proved. □

Now let us introduce a concept, *group ring*, which is important for Hajós' proof. Let $G = \{\mathbf{g}_1, \mathbf{g}_2, \ldots, \mathbf{g}_n\}$ be a finite abelian group under multiplication, then the group ring $\Re(G)$ is the set

$$\left\{\mathbf{r} = \sum z_i \mathbf{g}_i : \; z_i \in Z\right\}$$

in which the addition is defined by

$$\sum z_i \mathbf{g}_i + \sum z_i' \mathbf{g}_i = \sum (z_i + z_i') \mathbf{g}_i$$

and the multiplication is defined by

$$\left(\sum z_i \mathbf{g}_i\right) \cdot \left(\sum z_i' \mathbf{g}_i\right) = \sum \left( \sum_{\mathbf{g}_j \mathbf{g}_k = \mathbf{g}_i} z_j z_k' \right) \mathbf{g}_i,$$

where $\sum$ denotes a finite sum. For example, if $G = \{\mathbf{1}, \mathbf{a}\}$ with $|\mathbf{a}| = 2$, then

$$\Re(G) = \left\{ z_1 \mathbf{1} + z_2 \mathbf{a} : z_i \in Z \right\},$$

in which

$$(z_1 \mathbf{1} + z_2 \mathbf{a}) + (z_1' \mathbf{1} + z_2' \mathbf{a}) = (z_1 + z_1')\mathbf{1} + (z_2 + z_2')\mathbf{a}$$

and

$$(z_1 \mathbf{1} + z_2 \mathbf{a}) \cdot (z_1' \mathbf{1} + z_2' \mathbf{a}) = (z_1 z_1' + z_2 z_2')\mathbf{1} + (z_1 z_2' + z_1' z_2)\mathbf{a}.$$

Assume that

$$\mathbf{r}_i = \sum_{j=1}^{l_i} z_{ij} \mathbf{g}_{ij},$$

where $z_{ij} \neq 0$, then we define $G(\mathbf{r}_1, \ldots, \mathbf{r}_k)$ to be the subgroup of $G$ generated by $\{\mathbf{g}_{ij} : i = 1, \ldots, k;\ j = 1, \ldots, l_i\}$. In addition, if $p_1^{\alpha_1} p_2^{\alpha_2} \ldots p_m^{\alpha_m}$ is a standard factorization for $|G(\mathbf{r}_1, \ldots, \mathbf{r}_k)|$, then we define

$$p(\mathbf{r}_1, \ldots, \mathbf{r}_k) = \sum_{i=1}^{m} \alpha_i.$$

For convenience, for $\mathbf{g} \in G$, we use $\overline{\mathbf{g}}$ to denote either $\mathbf{1} - \mathbf{g}$ or $\mathbf{1} + \mathbf{g} + \cdots + \mathbf{g}^{p-1}$ for some prime number $p$. Clearly, $\overline{\mathbf{g}} \in \Re(G)$. Now we are ready to introduce another key lemma for Hajós' proof.

**Lemma 6.4 (Hajós, 1941).** *Assume that in the group ring $\Re(G)$ an equation*

$$\mathbf{r}\, \overline{\mathbf{g}_1} \cdots \overline{\mathbf{g}_k} = 0$$

*holds and none of the $\overline{\mathbf{g}_i}$ can be omitted without violating the equality. Then we have*

$$p(\mathbf{r}, \mathbf{g}_1, \ldots, \mathbf{g}_k) < p(\mathbf{r}) + k. \tag{6.8}$$

**Proof.** For convenience, we write

$$\mathbf{r} = \sum_{i=1}^{q} z_i \mathbf{h}_i,$$

where $z_i \neq 0$ and $\mathbf{h}_i \in G$. When $k = 1$, we proceed to show $\mathbf{g}_1 \in G(\mathbf{r})$ and hence

$$p(\mathbf{r}, \mathbf{g}_1) < p(\mathbf{r}) + 1.$$

If $\overline{\mathbf{g}_1} = \mathbf{1} - \mathbf{g}_1$, then it follows by $\mathbf{r} \cdot \overline{\mathbf{g}_1} = 0$ that

$$\sum_{i=1}^{q} z_i \mathbf{h}_i = \left( \sum_{i=1}^{q} z_i \mathbf{h}_i \right) \cdot \mathbf{g}_1$$

and therefore

$$\mathbf{h}_1 = \mathbf{h}_j \cdot \mathbf{g}_1$$

holds for some suitable $\mathbf{h}_j \in G$. Then we have $\mathbf{g}_1 = \mathbf{h}_1\mathbf{h}_j^{-1} \in G(\mathbf{r})$.

If $\overline{\mathbf{g}_1} = \mathbf{1} + \mathbf{g}_1 + \cdots + \mathbf{g}_1^{p-1}$, then we get $\mathbf{r}(\mathbf{1} - \mathbf{g}_1^p) = 0$. By repeating the above arguments we get $\mathbf{g}_1^p \in G(\mathbf{r})$. On the other hand, by the assumption we get

$$\mathbf{r} \cdot \left(\mathbf{g}_1 + \cdots + \mathbf{g}_1^{p-1}\right) = -\mathbf{r}$$

and therefore

$$\mathbf{h}_1 = \mathbf{h}_j \cdot \mathbf{g}_1^s$$

holds for some $j$ and $s$ with $1 \le s \le p-1$. Then we get $\mathbf{g}_1^s \in G(\mathbf{r})$. It is known in number theory that

$$up + vs = 1$$

holds for two integers $u$ and $v$. Thus we get

$$\mathbf{g}_1 = \mathbf{g}_1^{up+vs} = (\mathbf{g}_1^p)^u \cdot (\mathbf{g}_1^s)^v \in G(\mathbf{r}).$$

As a conclusion of the two cases, (6.8) is true for $k = 1$.

Inductively, assuming that $k \ge 2$ and

$$p(\mathbf{r}\,\overline{\mathbf{g}_1} \cdots \overline{\mathbf{g}_t}, \mathbf{g}_{t+1}, \ldots, \mathbf{g}_k) < p(\mathbf{r}\,\overline{\mathbf{g}_1} \cdots \overline{\mathbf{g}_t}) + k - t \tag{6.9}$$

holds for all $t = 1, \ldots, k-1$, we proceed to prove (6.8). Assume that $G_1$ and $G_2$ are two subgroups of $G$ such that $G_1 \subset G_2$ and $K$ is a subset of $G$. Let $H_i$ denote the subgroup generated by $\{G_i, K\}$; it is easy to see that the order of the quotient group $H_2/H_1$ is a factor of the order of the quotient group $G_2/G_1$. Thus we have

$$p(G_2, K) - p(G_1, K) \le p(G_2) - p(G_1).$$

Applying this inequality to $G_1 = G(\mathbf{r}\,\overline{g_1} \cdots \overline{\mathbf{g}_t})$, $G_2 = G(\mathbf{r}, \mathbf{g}_1, \ldots, \mathbf{g}_t)$, and $K = \{\mathbf{g}_{t+1}, \ldots, \mathbf{g}_k\}$, we get

$$\begin{aligned} p(\mathbf{r}, \mathbf{g}_1, \ldots, \mathbf{g}_k) &- p(\mathbf{r}\,\overline{\mathbf{g}_1} \cdots \overline{\mathbf{g}_t}, \mathbf{g}_{t+1}, \ldots, \mathbf{g}_k) \\ &\le p(\mathbf{r}, \mathbf{g}_1, \ldots, \mathbf{g}_t) - p(\mathbf{r}\,\overline{\mathbf{g}_1} \cdots \overline{\mathbf{g}_t}). \end{aligned}$$

Then it follows by (6.9) that

$$\begin{aligned} p(\mathbf{r}, \mathbf{g}_1, \ldots, \mathbf{g}_k) &- p(\mathbf{r}, \mathbf{g}_1, \ldots, \mathbf{g}_t) \\ &\le p(\mathbf{r}\,\overline{\mathbf{g}_1} \cdots \overline{\mathbf{g}_t}, \mathbf{g}_{t+1}, \ldots, \mathbf{g}_k) - p(\mathbf{r}\,\overline{\mathbf{g}_1} \cdots \overline{\mathbf{g}_t}) \\ &< k - t. \end{aligned} \tag{6.10}$$

Now we consider two cases.

**Case 1.** $p(\mathbf{g}_i)=1$ *holds for some* $\mathbf{g}_i$. Without loss of generality, we assume that $i=1$. Applying (6.10) with $t=1$, we get

$$p(\mathbf{r},\mathbf{g}_1,\ldots,\mathbf{g}_k)<k-1+p(\mathbf{r},\mathbf{g}_1)\le p(\mathbf{r})+k$$

and hence (6.8).

**Case 2.** $p(\mathbf{g}_i)\ge 2$ *holds for all* $\mathbf{g}_i$. In this case, we apply induction on $f=p(\mathbf{g}_1)+\cdots+p(\mathbf{g}_k)$. By multiplying $\overline{\mathbf{g}_k}$ with a suitable element $\mathbf{u}\in\Re(G)$, we can get a new element $\overline{\mathbf{g}_0}=\mathbf{1}-\mathbf{g}_0$ such that

$$1\le p(\mathbf{g}_0)=p(\mathbf{g}_k)-1.$$

To see this, we consider three subcases. For convenience, we assume that $|\mathbf{g}_k|=q^l m$, where $q$ is a prime and $(q,m)=1$.

**Subcase $\alpha$.** $\overline{\mathbf{g}_k}=\mathbf{1}-\mathbf{g}_k$. Then we can choose

$$\begin{cases}\mathbf{u}=\mathbf{1}+\mathbf{g}_k+\cdots+\mathbf{g}_k^{q-1}\\ \mathbf{g}_0=\mathbf{g}_k^q.\end{cases}$$

**Subcase $\beta$.** $\overline{\mathbf{g}_k}=\mathbf{1}+\mathbf{g}_k+\cdots+\mathbf{g}_k^{p-1}$ *and $p$ is a factor of $q^l m$*. Then we can choose

$$\begin{cases}\mathbf{u}=\mathbf{1}-\mathbf{g}_k\\ \mathbf{g}_0=\mathbf{g}_k^p.\end{cases}$$

**Subcase $\gamma$.** $\overline{\mathbf{g}_k}=\mathbf{1}+\mathbf{g}_k+\cdots+\mathbf{g}_k^{p-1}$ *and $p$ is not a factor of $q^l m$*. Then we can choose

$$\begin{cases}\mathbf{u}=(\mathbf{1}-\mathbf{g}_k)\left(\mathbf{1}+\mathbf{g}_k^p+\cdots+\mathbf{g}_k^{(q-1)p}\right)\\ \mathbf{g}_0=\mathbf{g}_k^{qp}.\end{cases}$$

Now we get a new equation

$$\mathbf{r}\,\overline{\mathbf{g}_0}\cdots\overline{\mathbf{g}_{k'}}=0,$$

where $k'\le k-1$ and

$$\sum_{i=0}^{k'}p(\mathbf{g}_i)<\sum_{i=1}^{k}p(\mathbf{g}_i).$$

Thus, by the inductive assumption on $f$, we get

$$p(\mathbf{r},\mathbf{g}_0,\ldots,\mathbf{g}_{k'})\le p(\mathbf{r})+k'. \tag{6.11}$$

If $k' = 0$, then it follows by (6.11) that $p(\mathbf{r}, \mathbf{g}_0) = p(\mathbf{r})$ and therefore $\mathbf{g}_0 \in G(\mathbf{r})$. By the construction of $\mathbf{g}_0$, we get

$$p(\mathbf{r}, \mathbf{g}_k) \leq p(\mathbf{r}) + 1.$$

Applying this inequality to (6.10), we get (6.8).

If $k' \geq 1$, it follows by (6.11) that

$$p(\mathbf{r}, \mathbf{g}_1, \ldots, \mathbf{g}_{k'}) \leq p(\mathbf{r}) + k'.$$

Then, taking $t = k'$ and applying this inequality to (6.10), we can deduce (6.8) as well. As a conclusion, the lemma is proved. □

**Proof of the conjecture.** Clearly the conjecture is true when $n = 1$. Assuming that the statement is true for $n \leq m - 1$, we proceed to show it for $n = m$.

Based on Lemma 6.3, we may assume that

$$G = [\mathbf{g}_1]_{p_1} \otimes \cdots \otimes [\mathbf{g}_m]_{p_m}, \tag{6.12}$$

where $p_i$ are prime numbers and none of the factors is a subgroup. Then we have $\mathbf{g}_m^{p_m} \neq \mathbf{1}$. It is clear that $\mathbf{g}_m^{p_m} \cdot G = G$, therefore we get

$$\mathbf{g}_m^{p_m} \cdot [\mathbf{g}_1]_{p_1} \otimes \cdots \otimes [\mathbf{g}_{m-1}]_{p_{m-1}} = [\mathbf{g}_1]_{p_1} \otimes \cdots \otimes [\mathbf{g}_{m-1}]_{p_{m-1}}. \tag{6.13}$$

Otherwise, if

$$\mathbf{g}_m^{p_m} \mathbf{g}_1^{s_1} \cdots \mathbf{g}_{m-1}^{s_{m-1}} = \mathbf{g}_m^{t_m} \mathbf{g}_1^{t_1} \cdots \mathbf{g}_{m-1}^{t_{m-1}}$$

holds for some $1 \leq t_m \leq p_m - 1$, $0 \leq s_i \leq p_i - 1$, and $0 \leq t_i \leq p_i - 1$, we can deduce

$$\mathbf{g}_m^{p_m - t_m} \mathbf{g}_1^{q(s_1 - t_1)} \cdots \mathbf{g}_{m-1}^{q(s_{m-1} - t_{m-1})} = \mathbf{g}_1^{q(t_1 - s_1)} \cdots \mathbf{g}_{m-1}^{q(t_{m-1} - s_{m-1})},$$

where

$$q(x) = \begin{cases} x & \text{if } x > 0, \\ 0 & \text{otherwise,} \end{cases}$$

which contradicts the assumption (6.12). By deleting as many factors as possible from (6.13), we get a new equation, without loss of generality, say

$$\mathbf{g}_m^{p_m} \cdot [\mathbf{g}_1]_{p_1} \otimes \cdots \otimes [\mathbf{g}_w]_{p_w} = [\mathbf{g}_1]_{p_1} \otimes \cdots \otimes [\mathbf{g}_w]_{p_w} \tag{6.14}$$

with $1 \leq w \leq m - 1$. For convenience, we write

$$W = [\mathbf{g}_1]_{p_1} \otimes \cdots \otimes [\mathbf{g}_w]_{p_w}$$

and

$$H = G(\mathbf{g}_1, \ldots, \mathbf{g}_w).$$

For every $\mathbf{h} \in H$, we have

$$\mathbf{h} = \mathbf{h}_1 \cdot \mathbf{h}_2,$$

where $\mathbf{h}_1 \in W$ and $\mathbf{h}_2 \in [\mathbf{g}_{w+1}]_{p_{w+1}} \otimes \cdots \otimes [\mathbf{g}_m]_{p_m}$. Since both $\mathbf{h}_1$ and $\mathbf{h}_2$ belong to $H$, we get

$$\mathbf{h}_2 \cdot W \subseteq H$$

and therefore $H$ can be divided into disjoint sets of the form $\mathbf{h}_2 \cdot W$. As a consequence, we get card$\{W\}$ divides card$\{H\}$; that is

$$p_1 \cdots p_w \mid \operatorname{card}\{H\}. \tag{6.15}$$

Let us define

$$\wp = \sum_{\mathbf{g} \in W} \mathbf{g}$$

and

$$\overline{\mathbf{g}_i} = \mathbf{1} + \mathbf{g}_i + \cdots + \mathbf{g}_i^{p_i - 1}$$

for $i = 1, 2, \ldots, w$. It is easy to see that

$$\wp = \prod_{i=1}^{w} \overline{\mathbf{g}_i}.$$

On the other hand, by (6.14) we get

$$\mathbf{g}_m^{p_m} \overline{\mathbf{g}_1} \cdots \overline{\mathbf{g}_w} = \overline{\mathbf{g}_1} \cdots \overline{\mathbf{g}_w}$$

and thus

$$\overline{\mathbf{g}_1} \cdots \overline{\mathbf{g}_w} \, (\mathbf{1} - \mathbf{g}_m^{p_m}) = 0.$$

Apply Lemma 6.4 with $\mathbf{r} = \mathbf{1}$, $k = w + 1$, and $\mathbf{g}_{w+1} = \mathbf{g}_m^{p_m}$, we get

$$p(\mathbf{g}_1, \ldots, \mathbf{g}_w, \mathbf{g}_{w+1}) < w + 1$$

and thus

$$p(\mathbf{g}_1, \ldots, \mathbf{g}_w) \leq w. \tag{6.16}$$

By (6.15) and (6.16), we get

$$G(\mathbf{g}_1, \ldots, \mathbf{g}_w) = [\mathbf{g}_1]_{p_1} \otimes \cdots \otimes [\mathbf{g}_w]_{p_w},$$

where $w \leq m - 1$. The conjecture follows by the inductive assumption. □

Now let us restate the proved conjecture as the following theorem.

**The Minkowski–Hajós theorem.** *Every lattice tiling $I^n + \Lambda$ of $E^n$ has a twin.*

## 6.4 Other versions

Let $I^n + \Lambda$ be a lattice tiling of $E^n$. By the Minkowski–Hajós theorem, without loss of generality, we may assume that $(1, 0, \ldots, 0) \in \Lambda$. Then we have $(k, 0, \ldots, 0) \in \Lambda$ for any integer $k$ and $\cup_{k=-\infty}^{\infty}(I^n + (k, 0, \ldots, 0))$ is a cylinder of infinite length. Let $H$ denote the hyperplane $\{\mathbf{x} \in E^n : x_1 = 0\}$, let $\Lambda'$ denote the projection of $\Lambda$ to $H$, and let $I'$ denote the projection of the cylinder to $H$. It is clear that $I'$ is an $(n-1)$-dimensional unit cube, $\Lambda'$ is an $(n-1)$-dimensional lattice, and $I' + \Lambda'$ is a tiling of $H$. By repeating this argument, one can easily deduce the following corollary.

**Corollary 6.1.** *If $I^n + Z^n A$ is a lattice tiling of $E^n$, after a suitable permutation of the axes, we can take*

$$A = \begin{pmatrix} 1 & 0 & \cdots & 0 \\ \alpha_{21} & 1 & \cdots & 0 \\ \vdots & \vdots & \ddots & \vdots \\ \alpha_{n1} & \alpha_{n2} & \cdots & 1 \end{pmatrix}.$$

Let $C$ be an $n$-dimensional centrally symmetric convex body and let $\Lambda = Z^n A$ be a lattice with a basis $\{\mathbf{a}_1, \mathbf{a}_2, \ldots, \mathbf{a}_n\}$, where $\mathbf{a}_i = (a_{i1}, a_{i2}, \ldots, a_{in})$. According to Minkowski's first theorem in geometry of numbers, *if*

$$v(C) \geq 2^n d(\Lambda) = 2^n \det(A),$$

*then $C$ contains a lattice point $\mathbf{u} \in \Lambda \setminus \{\mathbf{o}\}$.* As a special case $C = 2I^n$, we can easily deduce that *for any $n \times n$ real matrix $A = (a_{ij})$ with* $\det(A) = 1$ *there is a $\mathbf{z} \in Z^n \setminus \{\mathbf{o}\}$ satisfying*

$$\begin{cases} |a_{11}z_1 + a_{21}z_2 + \cdots + a_{n1}z_n| \leq 1 \\ |a_{12}z_1 + a_{22}z_2 + \cdots + a_{n2}z_n| \leq 1 \\ \qquad \cdots \\ |a_{1n}z_1 + a_{2n}z_2 + \cdots + a_{nn}z_n| \leq 1. \end{cases} \tag{6.17}$$

This is a basic result in Diophantine approximation. Therefore it is both important and interesting to study the equality cases.

In 1896 Minkowski discovered that, *for any $2 \times 2$ real matrix $A = (a_{ij})$ with* $\det(A) = 1$, *there is a $\mathbf{z} \in Z^2 \setminus \{\mathbf{o}\}$ satisfying*

$$\begin{cases} |a_{11}z_1 + a_{21}z_2| < 1 \\ |a_{12}z_1 + a_{22}z_2| < 1, \end{cases}$$

*unless $A$ has an integral column.* In that book he also stated an $n$-dimensional analogue and promised a proof, which has never appeared. In fact, the $n$-dimensional analogue is nothing else but a Diophantine approximation version of Minkowski's conjecture.

*If $A = (a_{ij})$ is an $n \times n$ real matrix with $\det(A) = 1$, then there is a $\mathbf{z} \in Z^n \setminus \{\mathbf{o}\}$ satisfying*

$$\begin{cases} |a_{11}z_1 + a_{21}z_2 + \cdots + a_{n1}z_n| < 1 \\ |a_{12}z_1 + a_{22}z_2 + \cdots + a_{n2}z_n| < 1 \\ \cdots \\ |a_{1n}z_1 + a_{2n}z_2 + \cdots + a_{nn}z_n| < 1, \end{cases} \tag{6.18}$$

*unless $A$ has an integral column.*

In fact, this is the first version. The geometric version was formulated in Minkowski (1907) 11 years later.

Next we will show their equivalence. It is easy to see that $I^n + \Lambda$ is a packing in $E^n$ if and only if (6.18) has no nontrivial integer solution. In addition, since

$$\frac{v(I^n)}{d(\Lambda)} = \frac{v(I^n)}{\det(A)} = 1,$$

$I^n + \Lambda$ will be a tiling of $E^n$ if it is a packing.

If the Diophantine approximation version is right, then $A$ has an integral column satisfying $(a_{1i}, a_{2i}, \ldots, a_{ni}) = 1$ whenever $I^n + Z^n A$ is a lattice tiling of $E^n$. Assume that $z_1, z_2, \ldots, z_n$ are integers satisfying

$$a_{1i}z_1 + a_{2i}z_2 + \cdots + a_{ni}z_n = 1.$$

We define

$$\mathbf{u} = \sum_{j=1}^{n} z_j \mathbf{a}_j,$$

$$U = \bigcup_{z \in Z} \left( I^n + z\mathbf{u} \right),$$

and

$$H = \left\{ \mathbf{x} \in E^n : \ x_i = 0 \right\}.$$

Clearly, $I^n + \Lambda$ consists of translates of $U$ and the intersection of these translates with $H$ is an $(n-1)$-dimensional lattice cube tiling of $H$. By an inductive argument, we can deduce the geometric version. On the other hand, the Diophantine approximation version follows directly from the Minkowski–Hajós theorem.

In 1998, Kolountzakis stated another version of this conjecture.

*Let $A = (a_{ij})$ be an $n \times n$ real matrix with $det(A) = 1$. If, for all $z \in Z^n \setminus \{\mathbf{o}\}$, some coordinate of the vector $zA$ is a nonzero integer, then $A$ has an integral column.*

We will not show their equivalence here. Instead, let us end this chapter with a conjecture about a general tile $C$. Just like the cube case, we call $C + \mathbf{x}$ and $C + \mathbf{y}$ a twin if they join at a whole facet.

**Conjecture 6.1.** *Every lattice tiling $C + \Lambda$ of $E^n$ has a twin.*

# 7

# Furtwängler's conjecture

## 7.1 Furtwängler's conjecture

Let $I^n + X$ be a set of unit cubes. If every point of $E^n$, which is not on the boundary of any cube, lies in exactly $k$ cubes, then we say that $I^n + X$ is a *k-fold tiling* of $E^n$. If at the same time $X$ is a lattice, then we say that $I^n + X$ is a *k-fold lattice tiling* of $E^n$. Especially, a tiling is a 1-fold tiling and a lattice tiling is a 1-fold lattice tiling of $E^n$.

By overlapping a given tiling exactly $k$ times, we can get a $k$-fold tiling of $E^n$. Clearly, the multiple tilings of this kind cannot reflect the nature of their complexity. The following example does show another class of multiple tilings.

**Example 7.1.** Let $\overline{I^n} + \Lambda$ be a tiling of $E^n$. Then, for any positive integer $k$, the system $\overline{I^n} + \frac{1}{k}\Lambda$ is a $k^n$-fold lattice tiling of $E^n$.

This statement can be deduced by the following argument. First of all, $\frac{1}{k}\overline{I^n} + \frac{1}{k}\Lambda$ is a lattice tiling of $E^n$. Second, for any point $\mathbf{p}$, the system $\frac{1}{k}\overline{I^n} + \mathbf{p} + \frac{1}{k}\Lambda$ is a tiling of $E^n$. Finally, there are $k^n$ points $\mathbf{p}_1, \mathbf{p}_2, \ldots, \mathbf{p}_{k^n}$ such that

$$\left(\text{int}\left(\frac{1}{k}\overline{I^n}\right) + \mathbf{p}_i\right) \bigcap \left(\text{int}\left(\frac{1}{k}\overline{I^n}\right) + \mathbf{p}_j\right) = \emptyset$$

holds for every pair of distinct points $\mathbf{p}_i$ and $\mathbf{p}_j$, and

$$\overline{I^n} = \bigcup_{i=1}^{k^n} \left(\frac{1}{k}\overline{I^n} + \mathbf{p}_i\right).$$

Therefore $\overline{I^n} + \frac{1}{k}\Lambda$ is a $k^n$-fold lattice tiling of $E^n$.

**Remark 7.1.** In fact, by a similar argument, we can show that, for any $n$-dimensional tile $C$, if $C + \Lambda$ is a lattice tiling of $E^n$ and $k$ is a positive integer, then $C + \frac{1}{k}\Lambda$ is a $k^n$-fold lattice tiling of $E^n$.

It is clear that the multiple lattice tiling is a very natural generalization for the lattice tiling. In 1936, even before Hajós' proof for Minkowski's conjecture, Furtwängler made the following generalization and did prove the $n \leq 3$ cases.

**Furtwängler's conjecture.** *In every multiple lattice tiling* $I^n + \Lambda$ *there is a twin.*

A few years later, without knowledge of Furtwängler's work, G. Hajós did study the same problem. First of all, with the very same argument as the lattice tiling case, he was able to restate the problem into the following algebraic version. Then he proved the $n \leq 3$ cases and discovered counterexamples for the other cases.

**Furtwängler's conjecture (Hajós, 1941).** *Let G be a finite abelian group and let* $A_1, A_2, \ldots, A_n$ *be cyclic sets of G. If every element* $\mathbf{g} \in G$ *can be expressed in exact k distinct ways as*

$$\mathbf{g} = \mathbf{a}_1\mathbf{a}_2 \cdots \mathbf{a}_n$$

*with* $\mathbf{a}_i \in A_i$, *then one of the n cyclic sets is a subgroup of G.*

In this chapter, we will discuss this conjecture, especially the works of G. Hajós and R.M. Robinson. For convenience, let $O_k$ denote the multiplicative cyclic group $\{1, e^{2\pi i/k}, \ldots, e^{2(k-1)\pi i/k}\}$, where $i = \sqrt{-1}$. This notation will appear frequently in this chapter.

## 7.2 A theorem of Furtwängler and Hajós

In this section, we will prove Furtwängler's conjecture for $n \leq 3$. The two proofs discovered respectively by Furtwängler and by Hajós are very different in nature. In order to have a systematic and coherent chapter, we choose the algebraic proof which belongs to Hajós.

**Theorem 7.1 (Furtwängler, 1936 and Hajós, 1941).** *Let k be a fixed positive integer, let G be a finite abelian group, and let* $A_1$, $A_2$, *and* $A_3$ *be three cyclic sets of G. If every element* $\mathbf{g} \in G$ *can be expressed in exact k distinct ways as*

$$\mathbf{g} = \mathbf{a}_1\mathbf{a}_2\mathbf{a}_3$$

*with* $\mathbf{a}_i \in A_i$, *then one of the three cyclic sets is a subgroup of G.*

**Proof.** Let $\mathbf{1}$ be the unit of $G$ and let $\Re(G)$ denote the group ring generated by $G$. In addition, for a subset $A$ of $G$, we write

$$\widetilde{A} = \sum_{\mathbf{a}\in A} \mathbf{a}.$$

Then the assumption of Theorem 7.1 can be restated as

$$k\widetilde{G} = \widetilde{A}_1 \cdot \widetilde{A}_2 \cdot \widetilde{A}_3. \tag{7.1}$$

Assume that $A_i = \{\mathbf{1}, \mathbf{g}_i, \ldots, \mathbf{g}_i^{q_i-1}\}$, then we can deduce

$$k\widetilde{G}\cdot(\mathbf{1}-\mathbf{g}_i) = k\widetilde{G} - k\widetilde{G} = 0$$

and

$$\widetilde{A}_i \cdot (\mathbf{1}-\mathbf{g}_i) = \mathbf{1} - \mathbf{g}_i^{q_i}.$$

Therefore, multiplying both sides of (7.1) by $(\mathbf{1}-\mathbf{g}_1)(\mathbf{1}-\mathbf{g}_2)(\mathbf{1}-\mathbf{g}_3)$, we get

$$(\mathbf{1}-\mathbf{g}_1^{q_1})\cdot(\mathbf{1}-\mathbf{g}_2^{q_2})\cdot(\mathbf{1}-\mathbf{g}_3^{q_3}) = 0. \tag{7.2}$$

For convenience, we write $\mathbf{x}_i = \mathbf{g}_i^{q_i}$, then (7.2) can be rewritten as

$$\mathbf{x}_1 + \mathbf{x}_2 + \mathbf{x}_3 + \mathbf{x}_1\mathbf{x}_2\mathbf{x}_3 = \mathbf{1} + \mathbf{x}_1\mathbf{x}_2 + \mathbf{x}_2\mathbf{x}_3 + \mathbf{x}_1\mathbf{x}_3. \tag{7.3}$$

If $\mathbf{x}_i = \mathbf{1}$ holds for one of the three indices, then the theorem is proved. Assume that $\mathbf{x}_i \neq \mathbf{1}$ for $i = 1$, 2, and 3, then by (7.3) we get

$$\begin{cases} \mathbf{x}_1\mathbf{x}_2\mathbf{x}_3 = \mathbf{1} \\ \mathbf{x}_1\mathbf{x}_2 = \mathbf{x}_3 \\ \mathbf{x}_1\mathbf{x}_3 = \mathbf{x}_2 \\ \mathbf{x}_2\mathbf{x}_3 = \mathbf{x}_1 \end{cases} \tag{7.4}$$

and therefore

$$\mathbf{x}_1^2 = \mathbf{x}_2^2 = \mathbf{x}_3^2 = \mathbf{1}. \tag{7.5}$$

Multiplying the two sides of (7.1) by $(\mathbf{1}-\mathbf{g}_2)(\mathbf{1}-\mathbf{g}_3)$, we get

$$0 = \widetilde{A}_1 \cdot (\mathbf{1}-\mathbf{g}_2^{q_2})\cdot(\mathbf{1}-\mathbf{g}_3^{q_3}) = \widetilde{A}_1 \cdot (\mathbf{1}-\mathbf{x}_2)\cdot(\mathbf{1}-\mathbf{x}_3)$$

and therefore by (7.4)

$$\widetilde{A}_1 \cdot (\mathbf{x}_2 + \mathbf{x}_3) = \widetilde{A}_1 \cdot (\mathbf{1} + \mathbf{x}_2\mathbf{x}_3) = \widetilde{A}_1 \cdot (\mathbf{1} + \mathbf{x}_1). \tag{7.6}$$

Clearly every term on the right-hand side of (7.6) belongs to the cyclic group $\langle \mathbf{g}_1 \rangle$. Then every term on the left-hand side also belongs to $\langle \mathbf{g}_1 \rangle$ and thus both $\mathbf{x}_2$ and $\mathbf{x}_3$ belong to $\langle \mathbf{g}_1 \rangle$.

If $|\mathbf{g}_1|$ is odd, then $\mathbf{x}^2 = \mathbf{1}$ has only one solution $\mathbf{x} = \mathbf{1}$ in $\langle \mathbf{g}_1 \rangle$. Then by (7.5) we get

$$\mathbf{x}_1 = \mathbf{x}_2 = \mathbf{x}_3 = \mathbf{1}.$$

The theorem is true. If $|\mathbf{g}_1| = 2l$ is even, then $\mathbf{x}^2 = \mathbf{1}$ has two solutions $\mathbf{x} = \mathbf{1}$ and $\mathbf{x} = \mathbf{g}_1^l$ in $\langle \mathbf{g}_1 \rangle$. Then by (7.4) and (7.5) we get

$$\mathbf{x}_i = \mathbf{1}$$

for one of the three indices. In this case, the theorem is also true. As a conclusion, the theorem is proved. □

By a similar but much simpler argument, we can prove the case of two cyclic sets as well. Therefore, we get the following theorem.

**Theorem 7.1***. *When $n \le 3$, every multiple lattice tiling $I^n + \Lambda$ has a twin.*

## 7.3 Hajós' counterexamples

**Theorem 7.2 (Hajós, 1938 and 1941).** *Whenever $n \ge 4$, Furtwängler's conjecture is not true for some $k$.*

**Proof.** First, we claim that, if the conjecture is not true for some $n$ and $k$, then it is false for $n+1$ and $2k$. Assume that

$$k\widetilde{G} = \widetilde{A_1} \cdot \widetilde{A_2} \cdots \widetilde{A_n} \tag{7.7}$$

holds with $n$ cyclic sets $A_i$, none of which is a subgroup. If

$$A_1 = \left\{\mathbf{1}, \mathbf{g}_1, \ldots, \mathbf{g}_1^{q_1-1}\right\},$$

it is easy to see that $A_{n+1} = \{\mathbf{1}, \mathbf{g}_1\}$ is a cyclic set but not a subgroup. Then by (7.7) we get

$$\widetilde{A_1} \cdot \widetilde{A_2} \cdots \widetilde{A_{n+1}} = k\widetilde{G} \cdot (\mathbf{1} + \mathbf{g}_1) = 2k\widetilde{G},$$

which implies the claim.

Second, we claim that the conjecture is false for $n = 4$ and $k = 9$. Let us take $G = O_{12} \otimes O_2$ and write $\mathbf{g}_1 = (e^{\pi i/6}, 1)$ and $\mathbf{g}_2 = (1, e^{\pi i})$. Clearly, $|\mathbf{g}_1| = 12$ and $|\mathbf{g}_2| = 2$. Then we define

$$A_1 = \left\{\mathbf{1}, \mathbf{g}_1, \ldots, \mathbf{g}_1^5\right\},$$

$$A_2 = \left\{\mathbf{1}, \mathbf{g}_1\mathbf{g}_2, (\mathbf{g}_1\mathbf{g}_2)^2, (\mathbf{g}_1\mathbf{g}_2)^3\right\},$$

$$A_3 = \left\{\mathbf{1}, \mathbf{g}_1^2\mathbf{g}_2, (\mathbf{g}_1^2\mathbf{g}_2)^2\right\},$$

and

$$A_4 = \left\{ \mathbf{1},\ \mathbf{g}_1^4\mathbf{g}_2,\ (\mathbf{g}_1^4\mathbf{g}_2)^2 \right\}.$$

It is easy to see that all $A_1$, $A_2$, $A_3$, and $A_4$ are cyclic sets, and none of them is a subgroup of $G$. Then, by a routine computation, we can deduce

$$\widetilde{A}_1 \cdot \widetilde{A}_2 \cdot \widetilde{A}_3 \cdot \widetilde{A}_4 = 9\widetilde{G}.$$

As a conclusion, the theorem is proved. □

Theorem 7.2 can be restated in the following geometric form.

**Theorem 7.2*.** *Whenever $n \geq 4$, there is a multiple lattice tiling of $E^n$, which has no twin.*

Geometrically speaking, the example constructed in the proof is corresponding to the nine-fold lattice tiling $I^4 + \Lambda$ of $E^4$, where $\Lambda$ is the lattice generated by $\mathbf{a}_1 = (2, 0, 0, 0)$, $\mathbf{a}_2 = \left(\frac{1}{3}, -\frac{1}{2}, 0, 0\right)$, $\mathbf{a}_3 = \left(\frac{1}{6}, \frac{1}{4}, \frac{1}{3}, 0\right)$, and $\mathbf{a}_4 = \left(\frac{1}{2}, \frac{1}{4}, 0, \frac{1}{3}\right)$. Once we get an $n$-dimensional example, by considering layers we can easily extend it to $(n+1)$ dimensions.

**Remark 7.2.** The geometric example was discovered by Hajós in 1938. Three years later he published the algebraic version in his famous paper (Hajós, 1941).

## 7.4 Robinson's characterization

To improve Theorem 7.2, in 1979 R.M. Robinson was able to determine all the integer pairs $\{n, k\}$ for which Furtwängler's conjecture is false.

**Theorem 7.3 (Robinson, 1979).** *There is a $k$-fold lattice tiling $I^n + \Lambda$ of $E^n$, which has no twin if and only if*

1 $n = 4$ *and $k$ is divisible by the square of an odd prime.*
2 $n = 5$ *and $k = 3$ or $k \geq 5$.*
3 $n \geq 6$ *and $k \geq 2$.*

This theorem can be restated in an algebraic version as follows.

**Theorem 7.3*.** *The group ring equation*

$$k\widetilde{G} = \widetilde{A}_1 \cdot \widetilde{A}_2 \cdots \widetilde{A}_n \tag{7.8}$$

*has a solution, where $A_i$ are cyclic subsets but not cyclic subgroups of $G$ if and only if*

1 $n = 4$ *and* $k$ *is divisible by the square of an odd prime.*
2 $n = 5$ *and* $k = 3$ *or* $k \geq 5$.
3 $n \geq 6$ *and* $k \geq 2$.

To prove this theorem, as we can imagine, we need several technical lemmas. For convenience, we write

$$\mathbf{h}_i = \mathbf{g}_i^{q_i}$$

and

$$\overline{A_i} = \widetilde{A}_i \cdot (\mathbf{g}_i - \mathbf{1}) = \mathbf{g}_i^{q_i} - \mathbf{1} = \mathbf{h}_i - \mathbf{1}.$$

Then it follows by (7.8) that

$$\overline{A_1} \cdot \widetilde{A}_2 \cdots \widetilde{A}_n = (\mathbf{g}_1 - \mathbf{1}) \prod_{i=1}^{n} \widetilde{A}_i = k\mathbf{g}_1 \cdot \widetilde{G} - k\widetilde{G} = 0$$

and, similarly

$$\prod_{i \in I} \overline{A_i} \cdot \prod_{i \in I'} \widetilde{A}_i = 0 \tag{7.9}$$

whenever $I$ is a nonempty subset of $\{1, 2, \ldots, n\}$ and

$$I' = \{1, 2, \ldots, n\} \setminus I.$$

Especially, when $I = \{1, 2, \ldots, n\}$, we have

$$\prod_{i=1}^{n} \overline{A_i} = \prod_{i=1}^{n} (\mathbf{h}_i - \mathbf{1}) = 0. \tag{7.10}$$

We will call $\{\mathbf{h}_1, \mathbf{h}_2, \ldots, \mathbf{h}_n\}$ a *primitive solution* to (7.10) if none of the factors can be omitted and will call it a solution of type $(|\mathbf{h}_1|, |\mathbf{h}_2|, \ldots, |\mathbf{h}_n|)$. For convenience, we assume that $|\mathbf{h}_i|$ are nondecreasing; that is

$$|\mathbf{h}_1| \leq |\mathbf{h}_2| \leq \cdots \leq |\mathbf{h}_n|.$$

Of course, every solution of (7.10) contains a primitive solution of some equation

$$\prod_{i \in I} (\mathbf{h}_i - \mathbf{1}) = 0,$$

where $I$ is a subset of $\{1, 2, \ldots, n\}$. In fact, Furtwängler's conjecture predicts that $\mathbf{h}_i - \mathbf{1} = 0$ holds at least for one of the $n$ indices.

**Lemma 7.1 (Robinson, 1979).** *If* $n \geq 3$ *and* $\{\mathbf{h}_1, \mathbf{h}_2, \ldots, \mathbf{h}_n\}$ *is a primitive solution to* (7.10), *then we have* $|\mathbf{h}_i| \leq 2^{n-2}$ *for all i. Especially, if both n and* $|\mathbf{h}_i|$ *are odd, then* $|\mathbf{h}_i| \leq 2^{n-3}$.

**Proof.** Without loss of generality, we only consider $\mathbf{h}_n$. By a routine computation, we have

$$\prod_{i=1}^{n-1} \overline{A_i} = \prod_{i=1}^{n-1} (\mathbf{h}_i - \mathbf{1}) = U_0 - V_0 = U - V \neq 0, \tag{7.11}$$

where we first multiply out, collect terms with same signs, and then cancel all the common terms. It is easy to see that both $U_0$ and $V_0$ have $2^{n-2}$ terms and not all of the terms are cancelled.

By (7.10) we get

$$(U - V) \cdot (\mathbf{h}_n - \mathbf{1}) = \mathbf{h}_n \cdot U + V - (\mathbf{h}_n \cdot V + U) = 0$$

and therefore

$$\mathbf{h}_n \cdot U + V = \mathbf{h}_n \cdot V + U.$$

Since $U$ and $V$ have no common term, it follows that

$$\begin{cases} \mathbf{h}_n \cdot U = U \\ \mathbf{h}_n \cdot V = V. \end{cases} \tag{7.12}$$

This means that the multiplication by $\mathbf{h}_n$ permutes the terms of $U$ in cycles of length $|\mathbf{h}_n|$ and therefore $|\mathbf{h}_n|$ is a divisor of the number of the terms in $U$. Thus we have

$$|\mathbf{h}_n| \leq |U| \leq |U_0| = 2^{n-2},$$

where $|U|$ denotes the number of the terms of $U$.

If $n$ is odd, then $n - 1$ is even. It follows by (7.11) that

$$\prod_{i \in I} \mathbf{h}_i \in U_0$$

if and only if

$$\prod_{i \notin I} \mathbf{h}_i \in U_0.$$

Thus the number of the terms in $U$ is even. Since $|\mathbf{h}_n|$ is an odd factor of $|U|$, which cannot exceed $2^{n-2}$, we get

$$|\mathbf{h}_n| \leq 2^{n-3}.$$

The lemma is proved. □

**Remark 7.3.** By repeating part of the argument of the previous proof, we can deduce that there is no primitive solution for (7.10) when $n = 2$. In other words, Furtwängler's conjecture is true in this case. It follows by (7.12) that $\mathbf{h}_n$ belongs to the subgroup of $G$ generated by $\mathbf{h}_1, \mathbf{h}_2, \ldots, \mathbf{h}_{n-1}$.

In the next three lemmas we will determine the possible types of the primitive solutions to (7.10) for $n = 3$, 4, and 5.

**Lemma 7.2 (Robinson, 1979).** *When $n = 3$, equation* (7.10) *has only primitive solutions of* $(2, 2, 2)$ *type.*

**Proof.** Let $U$ and $V$ be the elements of $\Re(G)$ defined by (7.11), it can be deduced that

$$\begin{cases} U = \mathbf{h}_1\mathbf{h}_2 + \mathbf{1} \\ V = \mathbf{h}_1 + \mathbf{h}_2. \end{cases}$$

Then, by applying (7.12), we get

$$\begin{cases} \mathbf{h}_1\mathbf{h}_2\mathbf{h}_3 = \mathbf{1} \\ \mathbf{h}_1\mathbf{h}_2 = \mathbf{h}_3 \\ \mathbf{h}_1\mathbf{h}_3 = \mathbf{h}_2 \\ \mathbf{h}_2\mathbf{h}_3 = \mathbf{h}_1 \end{cases}$$

and therefore

$$\mathbf{h}_1^2 = \mathbf{h}_2^2 = \mathbf{h}_3^2 = \mathbf{1}.$$

On the other hand, such a solution can be realized in the group ring $\Re(G)$ with $G = O_2 \otimes O_2$. Taking $\mathbf{h}_1 = (e^{\pi i}, 1)$, $\mathbf{h}_2 = (1, e^{\pi i})$, and $\mathbf{h}_3 = (e^{\pi i}, e^{\pi i})$, it can be verified that $\{\mathbf{h}_1, \mathbf{h}_2, \mathbf{h}_3\}$ is a primitive solution to (7.10). The lemma is proved. □

**Lemma 7.3 (Robinson, 1979).** *When $n = 4$, equation* (7.10) *has and only has primitive solutions of* $(3, 3, 3, 3)$ *type and* $(2, 2, 4, 4)$ *type.*

**Proof.** Assume that $\{\mathbf{h}_1, \mathbf{h}_2, \mathbf{h}_3, \mathbf{h}_4\}$ is a primitive solution to (7.10). It follows by

$$\overline{A_1} \cdot \overline{A_2} \cdot \overline{A_3} = (\mathbf{h}_1 - \mathbf{1})(\mathbf{h}_2 - \mathbf{1})(\mathbf{h}_3 - \mathbf{1}) = U_0 - V_0$$

that

$$\begin{cases} U_0 = \mathbf{h}_1\mathbf{h}_2\mathbf{h}_3 + \mathbf{h}_1 + \mathbf{h}_2 + \mathbf{h}_3 \\ V_0 = \mathbf{h}_1\mathbf{h}_2 + \mathbf{h}_1\mathbf{h}_3 + \mathbf{h}_2\mathbf{h}_3 + \mathbf{1}. \end{cases}$$

Then the only pairs of terms which might cancel out are among $\{\mathbf{h}_1\mathbf{h}_2\mathbf{h}_3,\ \mathbf{1}\}$, $\{\mathbf{h}_1,\ \mathbf{h}_2\mathbf{h}_3\}$, $\{\mathbf{h}_2,\ \mathbf{h}_1\mathbf{h}_3\}$, and $\{\mathbf{h}_3,\ \mathbf{h}_1\mathbf{h}_2\}$. Now we consider three cases.

**Case 1.** $|\mathbf{h}_4| = 2$. Since $\{\mathbf{h}_1, \mathbf{h}_2, \mathbf{h}_3, \mathbf{h}_4\}$ is primitive, by the assumption that $|\mathbf{h}_i|$ are nondecreasing we have $|\mathbf{h}_i| = 2$ for all $i$. Thus, if any pair of the terms are cancelled, then all the terms would be cancelled. So we have $U = U_0$ and $V = V_0$. By (7.12), without loss of generality, we get

$$\mathbf{h}_1\mathbf{h}_2\mathbf{h}_3\mathbf{h}_4 = \mathbf{h}_1$$

and therefore

$$\mathbf{h}_2\mathbf{h}_3\mathbf{h}_4 = \mathbf{1},$$

which leads to a solution of $(2, 2, 2)$ type on $\mathbf{h}_2$, $\mathbf{h}_3$, and $\mathbf{h}_4$. Hence there is no primitive solution in this case.

**Case 2.** $|\mathbf{h}_4| = 3$. In this case the multiplication by $\mathbf{h}_4$ to $U$ will permute three terms cyclically. Hence exactly one of the four pairs of terms must be cancelled between $U_0$ and $V_0$. Without loss of generality, we may suppose that $\mathbf{h}_1\mathbf{h}_2\mathbf{h}_3 = \mathbf{1}$ (if $\mathbf{h}_1 = \mathbf{h}_2\mathbf{h}_3$, we can take $\mathbf{h}_1' = \mathbf{h}_1^{-1}$ and get $\mathbf{h}_1'\mathbf{h}_2\mathbf{h}_3 = \mathbf{1}$) and, by (7.12)

$$\begin{cases} \mathbf{h}_1\mathbf{h}_4 = \mathbf{h}_2 \\ \mathbf{h}_2\mathbf{h}_4 = \mathbf{h}_3 \\ \mathbf{h}_3\mathbf{h}_4 = \mathbf{h}_1. \end{cases}$$

Then we get

$$\begin{aligned} \mathbf{h}_1\mathbf{h}_4 \cdot \mathbf{h}_1 &= \mathbf{h}_2 \cdot \mathbf{h}_3\mathbf{h}_4, \\ \mathbf{h}_1^2 &= \mathbf{h}_2\mathbf{h}_3, \\ \mathbf{h}_1^3 &= \mathbf{h}_1\mathbf{h}_2\mathbf{h}_3 = \mathbf{1}, \end{aligned}$$

and therefore

$$\mathbf{h}_1^3 = \mathbf{h}_2^3 = \mathbf{h}_3^3 = \mathbf{h}_4^3 = \mathbf{1}.$$

On the other hand, such a solution can be realized in the group ring $\mathfrak{R}(G)$ with $G = O_3 \otimes O_3$. To see this, we define $\mathbf{h}_1 = (e^{2\pi i/3}, 1)$, $\mathbf{h}_2 = (e^{2\pi i/3}, e^{2\pi i/3})$, $\mathbf{h}_3 = (e^{2\pi i/3}, e^{4\pi i/3})$, and $\mathbf{h}_4 = (1, e^{2\pi i/3})$. It can be verified that $\{\mathbf{h}_1, \mathbf{h}_2, \mathbf{h}_3, \mathbf{h}_4\}$ is a primitive solution to (7.10). Thus, in this case (7.10) has and only has a primitive solution of $(3, 3, 3, 3)$ type.

**Case 3.** $|\mathbf{h}_4| = 4$. In this case there is no cancellation between $U_0$ and $V_0$. Therefore, we have $U = U_0$ and $V = V_0$. Then the multiplication by $\mathbf{h}_4$ must permute the terms of $U$ and of $V$ cyclically. We may renumber the $\mathbf{h}_i$ so that the terms of $U$ occur in the order $\mathbf{h}_1\mathbf{h}_2\mathbf{h}_3$, $\mathbf{h}_1$, $\mathbf{h}_3$, $\mathbf{h}_2$. Then we get

$$\begin{cases} \mathbf{h}_2\mathbf{h}_3\mathbf{h}_4 = \mathbf{1} \\ \mathbf{h}_1\mathbf{h}_4 = \mathbf{h}_3 \\ \mathbf{h}_3\mathbf{h}_4 = \mathbf{h}_2 \\ \mathbf{h}_1\mathbf{h}_3 = \mathbf{h}_4 \end{cases}$$

and therefore

$$\begin{cases} \mathbf{h}_1^2 = \mathbf{h}_2^2 = \mathbf{1} \\ \mathbf{h}_3^4 = \mathbf{h}_4^4 = \mathbf{1}. \end{cases}$$

On the other hand, such a solution can be realized in the group ring $\mathfrak{R}(G)$ with $G = O_2 \otimes O_4$. Let us define $\mathbf{h}_1 = (e^{\pi i}, 1)$, $\mathbf{h}_2 = (e^{\pi i}, e^{\pi i})$, $\mathbf{h}_3 = (e^{\pi i}, e^{\pi i/2})$, and $\mathbf{h}_4 = (1, e^{\pi i/2})$. It can be verified that $\{\mathbf{h}_1, \mathbf{h}_2, \mathbf{h}_3, \mathbf{h}_4\}$ is a primitive solution to (7.10). Hence, in this case (7.10) has and only has a primitive solution of $(2, 2, 4, 4)$ type.

As a conclusion of these cases, the lemma is proved. □

**Lemma 7.4 (Robinson, 1979).** *If* $n = 5$ *and* $\{\mathbf{h}_1, \mathbf{h}_2, \ldots, \mathbf{h}_5\}$ *is a primitive solution to* (7.10), *then* $|\mathbf{h}_i| = 2$, 4, *or* 8.

**Proof.** For convenience, in this proof we remove the assumption that $|\mathbf{h}_i|$ are nondecreasing and only consider $|\mathbf{h}_5|$. Assume that

$$\begin{aligned} \overline{A_1} \cdot \overline{A_2} \cdot \overline{A_3} \cdot \overline{A_4} &= (\mathbf{h}_1 - \mathbf{1})(\mathbf{h}_2 - \mathbf{1})(\mathbf{h}_3 - \mathbf{1})(\mathbf{h}_4 - \mathbf{1}) \\ &= U_0 - V_0 = U - V. \end{aligned}$$

It was shown in the proof of Lemma 7.1 that $|\mathbf{h}_5|$ is a divisor of $|U|$ and $|U|$ is an even number less than or equal to 8. Thus the only doubtful case is $|U| = 6$. In this case we need to show that we cannot have $|\mathbf{h}_5| = 3$ or 6.

If $|\mathbf{h}_5| = 6$, clearly $\{\mathbf{h}_1, \mathbf{h}_2, \mathbf{h}_3, \mathbf{h}_4, \mathbf{h}_5^2\}$ is also a solution to (7.10). If it is primitive, we get a primitive solution with an element of order 3. If it is not primitive, then we can omit some factor $(\mathbf{h}_i - \mathbf{1})$ from (7.10) for some $i < 5$ and get a new equation. This can lead only to a primitive solution of type $(3, 3, 3, 3)$. Then the original solution $\{\mathbf{h}_1, \mathbf{h}_2, \ldots, \mathbf{h}_5\}$ would have three elements of order 3. Thus, if $|\mathbf{h}_5| = 6$ is possible, then $|\mathbf{h}_5| = 3$ is also possible. So it is sufficient to exclude the latter.

Assume that $|\mathbf{h}_5| = 3$, then we have

$$\begin{cases} U_0 = \mathbf{h}_1\mathbf{h}_2\mathbf{h}_3\mathbf{h}_4 + \mathbf{h}_1\mathbf{h}_2 + \mathbf{h}_1\mathbf{h}_3 + \mathbf{h}_1\mathbf{h}_4 + \mathbf{h}_2\mathbf{h}_3 + \mathbf{h}_2\mathbf{h}_4 + \mathbf{h}_3\mathbf{h}_4 + \mathbf{1} \\ V_0 = \mathbf{h}_1\mathbf{h}_2\mathbf{h}_3 + \mathbf{h}_1\mathbf{h}_2\mathbf{h}_4 + \mathbf{h}_1\mathbf{h}_3\mathbf{h}_4 + \mathbf{h}_2\mathbf{h}_3\mathbf{h}_4 + \mathbf{h}_1 + \mathbf{h}_2 + \mathbf{h}_3 + \mathbf{h}_4. \end{cases}$$

If $\mathbf{h}_1\mathbf{h}_2\mathbf{h}_3\mathbf{h}_4$ is cancelled by one of the first four terms of $V_0$, then one of the four elements $\mathbf{h}_1$, $\mathbf{h}_2$, $\mathbf{h}_3$, and $\mathbf{h}_4$ will be $\mathbf{1}$, which contradicts the assumption. If $\mathbf{h}_1\mathbf{h}_2\mathbf{h}_3\mathbf{h}_4$ is cancelled by one of the last four terms of $V_0$, then $\mathbf{1}$ will be cancelled by one of the first four terms of $V_0$. By similar arguments, we can conclude that the number of the cancelled terms of $U_0$ is even and therefore $|U| = 6$. By multiplying $\mathbf{h}_i^{-1}$ to every term, we can deduce that $\mathbf{1}$ plays the same role as any other term in $U_0$. Therefore, we may assume that $\mathbf{1}$ is one of the two cancelled terms in $U_0$. Since it has to be cancelled by one of the four terms $\mathbf{h}_1\mathbf{h}_2\mathbf{h}_3$, $\mathbf{h}_1\mathbf{h}_2\mathbf{h}_4$, $\mathbf{h}_1\mathbf{h}_3\mathbf{h}_4$, and $\mathbf{h}_2\mathbf{h}_3\mathbf{h}_4$ in $V_0$, the other term in $U_0$ to be cancelled must be $\mathbf{h}_1\mathbf{h}_2\mathbf{h}_3\mathbf{h}_4$. Therefore, we get

$$U = \mathbf{h}_1\mathbf{h}_2 + \mathbf{h}_1\mathbf{h}_3 + \mathbf{h}_1\mathbf{h}_4 + \mathbf{h}_2\mathbf{h}_3 + \mathbf{h}_2\mathbf{h}_4 + \mathbf{h}_3\mathbf{h}_4.$$

Since $\mathbf{h}_5 U = U$, the terms of $U$ must fall into two three-term cycles and one of them must contain at least two terms involving $\mathbf{h}_1$. We may suppose that the first cycle contains $\mathbf{h}_1\mathbf{h}_2$ and $\mathbf{h}_1\mathbf{h}_3$ in this order. Then we have $\mathbf{h}_2\mathbf{h}_5 = \mathbf{h}_3$ and $\mathbf{h}_2 = \mathbf{h}_3\mathbf{h}_5^2$. The second cycle contains at least one of the two terms $\mathbf{h}_2\mathbf{h}_4$ and $\mathbf{h}_3\mathbf{h}_4$, and hence also contains a term equal to the other. Exchanging two equal terms if necessary, we may suppose that both $\mathbf{h}_2\mathbf{h}_4$ and $\mathbf{h}_3\mathbf{h}_4$ occur in the second cycle. Without loss of generality, we assume that $\mathbf{h}_1\mathbf{h}_4$ belongs to the first cycle. Then we get

$$\begin{cases} \mathbf{h}_2\mathbf{h}_5 = \mathbf{h}_3 \\ \mathbf{h}_3\mathbf{h}_5 = \mathbf{h}_4 \\ \mathbf{h}_4\mathbf{h}_5 = \mathbf{h}_2. \end{cases} \tag{7.13}$$

Similarly, since $\mathbf{h}_5 V = V$, the terms of $V$ must fall into two three-term cycles. Applying (7.13) to $V$ and exchanging two equal terms if necessary, we may suppose that one of the two cycles is $\{\mathbf{h}_2, \mathbf{h}_3, \mathbf{h}_4\}$ and the other cycle contains the three complements $\mathbf{h}_1\mathbf{h}_2\mathbf{h}_3$, $\mathbf{h}_1\mathbf{h}_2\mathbf{h}_4$, and $\mathbf{h}_1\mathbf{h}_3\mathbf{h}_4$. It follows that the terms deleted from $V_0$ are $\mathbf{h}_1$ and $\mathbf{h}_2\mathbf{h}_3\mathbf{h}_4$. They must have cancelled $\mathbf{h}_1\mathbf{h}_2\mathbf{h}_3\mathbf{h}_4$ and $\mathbf{1}$ of $U_0$. Hence we have

$$\mathbf{h}_2\mathbf{h}_3\mathbf{h}_4 = \mathbf{1}.$$

However, this condition and (7.13) together imply that $\{\mathbf{h}_2, \mathbf{h}_3, \mathbf{h}_4, \mathbf{h}_5\}$ is a solution of $(3, 3, 3, 3)$ type. By this contradiction, the lemma is proved. □

Let $\chi(\mathbf{g})$ be a *homomorphic mapping* from $G$ to the multiplicative group of the *complex number field*. It is easy to see that $\chi(\mathbf{g})$ is an $m$th root of 1 for every $\mathbf{g} \in G$, if $|G| = m$. In fact, this function can be extended from $G$ to the group ring $\Re(G)$ by defining

$$\chi\left(\sum z_i \mathbf{g}_i\right) = \sum z_i \chi(\mathbf{g}_i).$$

This map, known as a *character* on $\Re(G)$, has some important properties. For example, there are exactly $m$ different characters. Let $\mathbf{g}_1 = \mathbf{1}, \mathbf{g}_2, \ldots, \mathbf{g}_m$ be the $m$ elements of $G$ and let $\chi_1, \chi_2, \ldots, \chi_m$ be the $m$ characters, where $\chi_1(\mathbf{g}) = 1$ holds for all $\mathbf{g} \in G$. Then we have

$$\sum_{j=1}^{m} \chi_i(\mathbf{g}_j) = 0 \qquad i \neq 1$$

and

$$\sum_{i=1}^{m} \chi_i(\mathbf{g}_j) = 0 \qquad j \neq 1.$$

Let

$$\mathbf{r} = \sum_{i=1}^{m} z_i \mathbf{g}_i$$

be an element in $\Re(G)$. It can be deduced from the above properties that

$$z_i = \frac{1}{m} \sum_{j=1}^{m} \chi_j(\mathbf{r})\, \chi_j(\mathbf{g}_i^{-1}).$$

Especially, if $\chi_j(\mathbf{r}) = 0$ for all $j$, then we get $z_i = 0$ for all $i$ and therefore $\mathbf{r} = 0$. If $\chi_j(\mathbf{r}) = 0$ for all $j \neq 1$, then we get $z_i = \chi_1(\mathbf{r})/m$, which is independent of $i$, and hence

$$\mathbf{r} = k\widetilde{G}$$

for some integer $k$.

**Lemma 7.5 (Robinson, 1979).** *If* $\{\mathbf{h}_1, \mathbf{h}_2, \ldots, \mathbf{h}_n\}$ *is a primitive solution to* (7.10) *and, for every i,* $|\mathbf{h}_i|$ *is a factor of* $|\mathbf{h}_n|$, *then*

$$\overline{A_1} \cdot \overline{A_2} \cdots \overline{A_{n-1}} \cdot \widetilde{A_n} \neq 0.$$

**Proof.** Based on the properties of the characters, it is sufficient to show that there is some character $\chi$ satisfying $\chi(\mathbf{h}_i) \neq 1$ for $i < n$, $\chi(\widetilde{A_n}) \neq 0$, and $\chi(\mathbf{h}_n) = \chi(\mathbf{g}_n)^{q_n} = 1$. Since

$$\begin{aligned} \chi(\widetilde{A_n}) &= 1 + \chi(\mathbf{g}_n) + \cdots + \chi(\mathbf{g}_n)^{q_n - 1} \\ &= \frac{\chi(\mathbf{h}_n) - 1}{\chi(\mathbf{g}_n) - 1}, \end{aligned}$$

we have $\chi(\widetilde{A_n}) = 0$ unless $\chi(\mathbf{g}_n) = 1$. Therefore it is sufficient to find a character $\chi$ satisfying

$$\begin{cases} \chi(\mathbf{h}_i) \neq 1 & \text{if } i < n, \\ \chi(\mathbf{h}_n) = \chi(\mathbf{g}_n) = 1. \end{cases} \tag{7.14}$$

Let $H$ denote the subgroup generated by $\{\mathbf{h}_1, \mathbf{h}_2, \ldots, \mathbf{h}_n\}$. We claim that, if there is such a character whenever $\mathbf{g}_n \in H$, then we can extend it into the general case. Let $\mathbf{g}$ be an element of $G$ such that $\mathbf{h}_n \in \langle \mathbf{g} \rangle$ and let $s$ be the smallest positive integer such that $\mathbf{g}^s \in H$. Since $\mathbf{h}_n \in \langle \mathbf{g} \rangle$ and $\mathbf{h}_n \in H$, we must have $\mathbf{h}_n \in \langle \mathbf{g}^s \rangle$. Based on the assumption, there is a character $\chi$ satisfying $\chi(\mathbf{h}_i) \neq 1$ if $i < n$ and $\chi(\mathbf{g}^s) = 1$. It is known in group theory that then we can extend the character from $H$ to the group generated by $H$ and $\mathbf{g}$ with $\chi(\mathbf{g}) = 1$.

Now we show the $\mathbf{g}_n \in H$ case. Since $|\mathbf{h}_i|$ is a factor of $|\mathbf{h}_n|$ for all $i$, we have

$$\mathbf{h}^{|\mathbf{h}_n|} = \mathbf{1}$$

for all $\mathbf{h} \in H$. Since $\mathbf{h}_n \in \langle \mathbf{g}_n \rangle$ and $\mathbf{g}_n^{|\mathbf{h}_n|} = \mathbf{1}$, we must have

$$\langle \mathbf{h}_n \rangle = \langle \mathbf{g}_n \rangle. \tag{7.15}$$

By the assumption that $\{\mathbf{h}_1, \mathbf{h}_2, \ldots, \mathbf{h}_n\}$ is a primitive solution to (7.10), it follows that there is a character $\chi$ on $H$ satisfying $\chi(\mathbf{h}_i) \neq 1$ for all $i < n$ and $\chi(\mathbf{h}_n) = 1$. Then by (7.15) we get $\chi(\mathbf{g}_n) = 1$ as well. As a conclusion we have shown (7.14). The lemma is proved. □

**Proof of Theorem 7.3, Case 1.** First, we deal with the necessary part. Suppose that

$$\begin{cases} \widetilde{A}_1 \cdot \widetilde{A}_2 \cdot \widetilde{A}_3 \cdot \widetilde{A}_4 = k\,\widetilde{G} \\ \mathbf{h}_i \neq \mathbf{1}, \qquad i = 1, 2, 3, 4 \end{cases} \tag{7.16}$$

holds for some positive integer $k$. Then we get

$$\begin{cases} \overline{A_1} \cdot \overline{A_2} \cdot \overline{A_3} \cdot \overline{A_4} = 0 \\ \overline{A_1} \cdot \overline{A_2} \cdot \overline{A_3} \cdot \widetilde{A}_4 = 0 \end{cases}$$

and therefore, by lemma 7.5

$$\overline{A_1} \cdot \overline{A_2} \cdot \overline{A_3} = 0.$$

By the proof of Lemma 7.2 we get

$$\begin{cases} \mathbf{h}_1\mathbf{h}_2\mathbf{h}_3 = \mathbf{1} \\ \mathbf{h}_1^2 = \mathbf{h}_2^2 = \mathbf{h}_3^2 = \mathbf{1}. \end{cases} \tag{7.17}$$

From these equations, we can deduce $\mathbf{h}_2 \notin \langle \mathbf{g}_1 \rangle$ and $\mathbf{h}_3 \notin \langle \mathbf{g}_1 \rangle$. Otherwise, since $\mathbf{x}^2 = \mathbf{1}$ has only two solutions in $\langle \mathbf{g}_1 \rangle$ and they are $\mathbf{x} = \mathbf{1}$ and $\mathbf{x} = \mathbf{h}_1$, we can deduce $\mathbf{h}_2 = \mathbf{h}_1$ and therefore $\mathbf{h}_3 = \mathbf{1}$.

By (7.16) we also have

$$\widetilde{A_1} \cdot \overline{A_2} \cdot \overline{A_3} \cdot \overline{A_4} = 0,$$

from which we can deduce

$$\widetilde{A_1}(\mathbf{h}_2\mathbf{h}_3\mathbf{h}_4 + \mathbf{h}_2 + \mathbf{h}_3 + \mathbf{h}_4) = \widetilde{A_1}(\mathbf{h}_2\mathbf{h}_3 + \mathbf{h}_2\mathbf{h}_4 + \mathbf{h}_3\mathbf{h}_4 + \mathbf{1}).$$

Then, applying (7.17), we can deduce

$$\widetilde{A_1}(\mathbf{1} + \mathbf{h}_1)(\mathbf{h}_2 + \mathbf{h}_4) = \widetilde{A_1}(\mathbf{1} + \mathbf{h}_1)(\mathbf{1} + \mathbf{h}_2\mathbf{h}_4).$$

Now the right-hand side, when expanded, contains powers of $\mathbf{g}_1$, hence the left-hand side must also. Since $\mathbf{h}_2 \notin \langle \mathbf{g}_1 \rangle$, we must have $\mathbf{h}_4 \in \langle \mathbf{g}_1 \rangle$. Similarly, by $\overline{A_1} \cdot \widetilde{A_2} \cdot \overline{A_3} \cdot \overline{A_4} = 0$ and $\overline{A_1} \cdot \overline{A_2} \cdot \widetilde{A_3} \cdot \overline{A_4} = 0$, we get $\mathbf{h}_4 \in \langle \mathbf{g}_2 \rangle$ and $\mathbf{h}_4 \in \langle \mathbf{g}_3 \rangle$.

Now we claim that $s = |\mathbf{h}_4|$ is odd. If, on the contrary, $s = 2t$, then we have $(\mathbf{h}_4^t)^2 = \mathbf{1}$ and $\mathbf{h}_4^t \in \langle \mathbf{g}_1 \rangle$. Since $\mathbf{h}_4^t \neq \mathbf{1}$, we must have $\mathbf{h}_4^t = \mathbf{h}_1$. Similarly, $\mathbf{h}_4^t = \mathbf{h}_2$. Hence $\mathbf{h}_1 = \mathbf{h}_2$, so that by (7.17) $\mathbf{h}_3 = \mathbf{1}$, which contradicts the hypothesis.

Since $\mathbf{h}_4 \in \langle \mathbf{g}_i \rangle$ for $i = 1, 2$, and 3, we have $s \mid |\mathbf{g}_i|$. On the other hand, by (7.17), we have $|\mathbf{g}_i| = 2q_i$ and therefore $s \mid q_i$. For convenience, we write $q_i = sd_i$. Furthermore, all solutions of $\mathbf{x}^s = \mathbf{1}$ in $\langle \mathbf{g}_i \rangle$ must be powers of $\mathbf{h}_4$, therefore $\mathbf{g}_i^{2d_i} \in \langle \mathbf{h}_4 \rangle$.

Writing

$$\begin{aligned} \widetilde{A_i}^* &= \widetilde{A_i}\left(\mathbf{1} + \mathbf{h}_i + \cdots + \mathbf{h}_i^{|\mathbf{h}_i|-1}\right) \\ &= \mathbf{1} + \mathbf{g}_i + \cdots + \mathbf{g}_i^{q_i|\mathbf{h}_i|-1}, \end{aligned}$$

since $|\mathbf{h}_i| = 2$ for $i = 1$, 2, and 3 and $|\mathbf{h}_4| = s$, it can be deduced by (7.16) that

$$\widetilde{A_1}^* \cdot \widetilde{A_2}^* \cdot \widetilde{A_3}^* \cdot \widetilde{A_4} = 2^3 k\, \widetilde{G}$$

and

$$\widetilde{A_1}^* \cdot \widetilde{A_2}^* \cdot \widetilde{A_3}^* \cdot \widetilde{A_4}^* = 2^3 sk\, \widetilde{G}. \tag{7.18}$$

On the other hand, since $q_i = sd_i$, for $i = 1$, 2, and 3, we have

$$\begin{aligned} \widetilde{A_i}^* &= \mathbf{1} + \mathbf{g}_i + \cdots + \mathbf{g}_i^{2sd_i-1} \\ &= \left(\mathbf{1} + \mathbf{g}_i + \cdots + \mathbf{g}_i^{2d_i-1}\right)\left(\mathbf{1} + \mathbf{g}_i^{2d_i} + \cdots + \mathbf{g}_i^{2(s-1)d_i}\right). \end{aligned} \tag{7.19}$$

It follows by the last paragraph that each term in the last factor is a power of $\mathbf{h}_4$ and hence of $\mathbf{g}_4$. Clearly, we have $\mathbf{g}_4^j \widetilde{A_4}^* = \widetilde{A_4}^*$. Thus by (7.18) and (7.19) we get

$$s^3 \prod_{i=1}^{3} \left(\mathbf{1} + \mathbf{g}_i + \cdots + \mathbf{g}_i^{2d_i - 1}\right) \widetilde{A_4}^* = 2^3 sk\, \widetilde{G}.$$

Since every term on the right-hand side has a coefficient $8sk$, it follows that $s^3 \mid 8sk$, or $s^2 \mid 8k$. Since $s$ is odd and $s > 1$, we conclude that $k$ is divisible by the square of an odd prime.

Now let us show the sufficient part. Assume that $p$ is an odd prime and $k$ is a positive integer. Let us take $G = O_{2p} \otimes O_4$ and write $\mathbf{e}_1 = (e^{\pi i/p}, 1)$ and $\mathbf{e}_2 = (1, e^{\pi i/2})$. Clearly, $|\mathbf{e}_1| = 2p$ and $|\mathbf{e}_2| = 4$. We define

$$\begin{aligned}
\widetilde{A_1} &= \mathbf{1} + \mathbf{e}_1 + \cdots + \mathbf{e}_1^{kp-1} \\
&= \left(\mathbf{1} + \mathbf{e}_1 + \cdots + \mathbf{e}_1^{p-1}\right)\left(\mathbf{1} + \mathbf{e}_1^{p} + \cdots + \mathbf{e}_1^{(k-1)p}\right), \\
\widetilde{A_2} &= \mathbf{1} + \mathbf{e}_1\mathbf{e}_2^2 + \cdots + (\mathbf{e}_1\mathbf{e}_2^2)^{p-1}, \\
\widetilde{A_3} &= \mathbf{1} + \mathbf{e}_1\mathbf{e}_2 + \cdots + (\mathbf{e}_1\mathbf{e}_2)^{2p-1},
\end{aligned}$$

and

$$\widetilde{A_4} = \mathbf{1} + \mathbf{e}_1^2\mathbf{e}_2 + \mathbf{e}_1^4\mathbf{e}_2^2 + \mathbf{e}_1^6\mathbf{e}_2^3.$$

Then it can be verified that

$$\begin{cases} \widetilde{A_1} \cdot \widetilde{A_2} \cdot \widetilde{A_3} \cdot \widetilde{A_4} = kp^2\, \widetilde{G} \\ \mathbf{h}_i \neq \mathbf{1}, \quad i = 1, 2, 3, 4. \end{cases}$$

The first case of the theorem is proved. □

**Proof of Theorem 7.3, Case 2.** First, we show the necessary part. Suppose that

$$\begin{cases} \widetilde{A_1} \cdot \widetilde{A_2} \cdot \widetilde{A_3} \cdot \widetilde{A_4} \cdot \widetilde{A_5} = k\, \widetilde{G} \\ \mathbf{h}_i \neq \mathbf{1}, \qquad i = 1, 2, 3, 4, 5 \end{cases} \tag{7.20}$$

holds with some positive integer $k$, then we get

$$\begin{cases} \overline{A_1} \cdot \overline{A_2} \cdot \overline{A_3} \cdot \overline{A_4} \cdot \overline{A_5} = 0 \\ \overline{A_1} \cdot \overline{A_2} \cdot \overline{A_3} \cdot \overline{A_4} \cdot \widetilde{A_5} = 0 \end{cases}$$

and therefore, by Lemma 7.5

$$\overline{A_1} \cdot \overline{A_2} \cdot \overline{A_3} \cdot \overline{A_4} = 0.$$

Thus, by Lemmas 7.2, 7.3, and 7.4, we only need to deal with the following subcases.

**Subcase 2.1.** $\overline{A_1} \cdot \overline{A_2} \cdot \overline{A_3} = 0$ *has a primitive solution of* $(2, 2, 2)$ *type.* First, we have

$$\begin{cases} \mathbf{h}_1\mathbf{h}_2\mathbf{h}_3 = \mathbf{1} \\ \mathbf{h}_1^2 = \mathbf{h}_2^2 = \mathbf{h}_3^2 = \mathbf{1} \end{cases} \tag{7.21}$$

and

$$\widetilde{A_1} \cdot \overline{A_2} \cdot \overline{A_3} \cdot \overline{A_4} \cdot \overline{A_5} = 0. \tag{7.22}$$

It follows by (7.21) that $\mathbf{h}_2\mathbf{h}_3 = \mathbf{h}_1$ and $\mathbf{h}_1\mathbf{h}_2 = \mathbf{h}_3$. Therefore, by (7.22), we get

$$\widetilde{A_1}(\mathbf{1}+\mathbf{h}_1)(\mathbf{1}+\mathbf{h}_2\mathbf{h}_4+\mathbf{h}_2\mathbf{h}_5+\mathbf{h}_4\mathbf{h}_5) = \widetilde{A_1}(\mathbf{1}+\mathbf{h}_1)(\mathbf{h}_2\mathbf{h}_4\mathbf{h}_5+\mathbf{h}_2+\mathbf{h}_4+\mathbf{h}_5). \tag{7.23}$$

When expanded, there will be powers of $\mathbf{g}_1$ on the left-hand side. Therefore, there must also be powers of $\mathbf{g}_1$ on the right-hand side, and thus $\mathbf{h}_2$, $\mathbf{h}_4$, $\mathbf{h}_5$ or $\mathbf{h}_2\mathbf{h}_4\mathbf{h}_5$ must be a power of $\mathbf{g}_1$. First of all, we cannot have $\mathbf{h}_2 \in \langle \mathbf{g}_1 \rangle$, since the only solutions for $\mathbf{x}^2 = \mathbf{1}$ in $\langle \mathbf{g}_1 \rangle$ are $\mathbf{x} = \mathbf{1}$ or $\mathbf{h}_1$, and $\mathbf{h}_2 = \mathbf{h}_1$ would lead to $\mathbf{h}_3 = \mathbf{1}$. Thus $\mathbf{h}_4$, $\mathbf{h}_5$ or $\mathbf{h}_2\mathbf{h}_4\mathbf{h}_5$ belongs to $\langle \mathbf{g}_1 \rangle$. In the latter case, we also have $\mathbf{h}_1\mathbf{h}_3\mathbf{h}_4\mathbf{h}_5 \in \langle \mathbf{g}_1 \rangle$ and therefore $\mathbf{h}_3\mathbf{h}_4\mathbf{h}_5 \in \langle \mathbf{g}_1 \rangle$. Since $\widetilde{A_1}(\mathbf{1}+\mathbf{h}_1)$ is cyclic and is unchanged by multiplying by a power of $\mathbf{g}_1$, we get

$$\widetilde{A_1}(\mathbf{1}+\mathbf{h}_1) = \widetilde{A_1}(\mathbf{1}+\mathbf{h}_1) \cdot \mathbf{h}_2\mathbf{h}_4\mathbf{h}_5,$$
$$\widetilde{A_1}(\mathbf{1}+\mathbf{h}_1) \cdot \mathbf{h}_4\mathbf{h}_5 = \widetilde{A_1}(\mathbf{1}+\mathbf{h}_1) \cdot \mathbf{h}_2$$

and

$$\widetilde{A_1}(\mathbf{1}+\mathbf{h}_1) \cdot \mathbf{h}_2 = \widetilde{A_1}(\mathbf{1}+\mathbf{h}_1) \cdot \mathbf{h}_4^{-1}\mathbf{h}_5^{-1}.$$

Then it can be deduced from (7.23) that

$$\widetilde{A_1}(\mathbf{1}+\mathbf{h}_1)(\mathbf{h}_4^{-1}+\mathbf{h}_5^{-1}) = \widetilde{A_1}(\mathbf{1}+\mathbf{h}_1)(\mathbf{h}_4+\mathbf{h}_5)$$

and hence

$$\widetilde{A_1}(\mathbf{1}+\mathbf{h}_1)(\mathbf{1}+\mathbf{h}_4\mathbf{h}_5^{-1}) = \widetilde{A_1}(\mathbf{1}+\mathbf{h}_1)(\mathbf{h}_4^2+\mathbf{h}_4\mathbf{h}_5). \tag{7.24}$$

Since $\mathbf{h}_2 \notin \langle \mathbf{g}_1 \rangle$, we get $\mathbf{h}_4\mathbf{h}_5 \notin \langle \mathbf{g}_1 \rangle$. By comparing the two sides of (7.24), we get $\mathbf{h}_4^2 \in \langle \mathbf{g}_1 \rangle$. Similarly, we find that $\mathbf{h}_5^2 \in \langle \mathbf{g}_1 \rangle$.

Summarizing these conclusions and similar conclusions drawn from the equations $\overline{A_1} \cdot \widetilde{A_2} \cdot \overline{A_3} \cdot \overline{A_4} \cdot \overline{A_5} = 0$ and $\overline{A_1} \cdot \overline{A_2} \cdot \widetilde{A_3} \cdot \overline{A_4} \cdot \overline{A_5} = 0$, we have

$$\begin{cases} \mathbf{h}_4 \in \langle \mathbf{g}_1 \rangle \text{ or } \mathbf{h}_5 \in \langle \mathbf{g}_1 \rangle \text{ or } \{\mathbf{h}_4^2, \mathbf{h}_5^2, \mathbf{h}_2\mathbf{h}_4\mathbf{h}_5, \mathbf{h}_3\mathbf{h}_4\mathbf{h}_5\} \subseteq \langle \mathbf{g}_1 \rangle \\ \mathbf{h}_4 \in \langle \mathbf{g}_2 \rangle \text{ or } \mathbf{h}_5 \in \langle \mathbf{g}_2 \rangle \text{ or } \{\mathbf{h}_4^2, \mathbf{h}_5^2, \mathbf{h}_1\mathbf{h}_4\mathbf{h}_5, \mathbf{h}_3\mathbf{h}_4\mathbf{h}_5\} \subseteq \langle \mathbf{g}_2 \rangle \\ \mathbf{h}_4 \in \langle \mathbf{g}_3 \rangle \text{ or } \mathbf{h}_5 \in \langle \mathbf{g}_3 \rangle \text{ or } \{\mathbf{h}_4^2, \mathbf{h}_5^2, \mathbf{h}_1\mathbf{h}_4\mathbf{h}_5, \mathbf{h}_2\mathbf{h}_4\mathbf{h}_5\} \subseteq \langle \mathbf{g}_3 \rangle. \end{cases} \tag{7.25}$$

Next we claim that it is impossible to have $\mathbf{h}_4^2 = \mathbf{h}_5^2 = \mathbf{1}$. If, on the contrary, these equations hold, we proceed to deduce a contradiction. Suppose that $\mathbf{h}_4$ or $\mathbf{h}_5$ belongs to some $\langle \mathbf{g}_i \rangle$ with $i \leq 3$. We may assume that $\mathbf{h}_4 \in \langle \mathbf{g}_1 \rangle$. It follows that $\mathbf{h}_4 = \mathbf{h}_1$. We cannot also have $\mathbf{h}_4 \in \langle \mathbf{g}_2 \rangle$, since this would lead to $\mathbf{h}_4 = \mathbf{h}_2$, and hence $\mathbf{h}_1 = \mathbf{h}_2$ and $\mathbf{h}_3 = \mathbf{1}$. Similarly, we get $\mathbf{h}_4 \notin \langle \mathbf{g}_3 \rangle$. Since $\mathbf{h}_1\mathbf{h}_4\mathbf{h}_5 = \mathbf{h}_5$, it follows by (7.25) that $\mathbf{h}_5 \in \langle \mathbf{g}_2 \rangle$ and $\mathbf{h}_5 \in \langle \mathbf{g}_3 \rangle$, which leads to a contradiction in the similar way. Thus we must have $\mathbf{h}_4 \notin \langle \mathbf{g}_i \rangle$ and $\mathbf{h}_5 \notin \langle \mathbf{g}_i \rangle$ for $i \leq 3$. Hence we must use the third alternative in each of the three relations in (7.25). Since $(\mathbf{h}_i\mathbf{h}_4\mathbf{h}_5)^2 = \mathbf{1}$, we have

$$\mathbf{h}_i\mathbf{h}_4\mathbf{h}_5 = \mathbf{1} \quad \text{or} \quad \mathbf{h}_i\mathbf{h}_4\mathbf{h}_5 = \mathbf{h}_j,$$

where $j \in \{1, 2, 3\} \setminus \{i\}$. Then we can deduce

$$\mathbf{h}_3\mathbf{h}_4\mathbf{h}_5 = \mathbf{h}_2\mathbf{h}_4\mathbf{h}_5 = \mathbf{1}$$

and hence $\mathbf{h}_2 = \mathbf{h}_3$ and $\mathbf{h}_1 = \mathbf{1}$, which contradicts the assumption.

Since we cannot have $\mathbf{h}_4^2 = \mathbf{h}_5^2 = \mathbf{1}$, we may assume that $\mathbf{h}_5^2 \neq \mathbf{1}$. Then it follows by (7.25) that

$$\mathbf{h}_4 \in \langle \mathbf{g}_i \rangle \quad \text{or} \quad \mathbf{h}_5^2 \in \langle \mathbf{g}_i \rangle$$

for $i \leq 3$. Thus one of the two alternatives must hold for at least two indices.

Suppose $\mathbf{h}_4 \in \langle \mathbf{g}_1 \rangle$ and $\mathbf{h}_4 \in \langle \mathbf{g}_2 \rangle$, repeating the proof arguments in Case 1 about $s$, we can get $s = |\mathbf{h}_4|$ is odd and $q_i = sd_i$ hold for some suitable integers $d_i$ for $i = 1$, 2, and 4. Then it follows from (7.20) that

$$\widetilde{A_1}^* \cdot \widetilde{A_2}^* \cdot \widetilde{A_3} \cdot \widetilde{A_4}^* \cdot \widetilde{A_5} = 2^2 sk\, \widetilde{G},$$

$$s^2 \prod_{i=1}^{2} \left(\mathbf{1} + \mathbf{g}_i + \cdots + \mathbf{g}_i^{2d_i - 1}\right) \cdot \widetilde{A_3} \cdot \widetilde{A_4}^* \cdot \widetilde{A_5} = 4sk\, \widetilde{G}$$

and therefore $k$ is divisible by an odd prime.

Suppose $\mathbf{h}_5^2 \in \langle \mathbf{g}_1 \rangle$ and $\mathbf{h}_5^2 \in \langle \mathbf{g}_2 \rangle$. We write $q_5' = 2q_5$, $\mathbf{h}_5' = \mathbf{g}_5^{q_5'} = \mathbf{h}_5^2$, and

$$\widetilde{A_5}' = \widetilde{A_5}(\mathbf{1} + \mathbf{h}_5) = \mathbf{1} + \mathbf{g}_5 + \cdots + \mathbf{g}_5^{q_5' - 1},$$

it follows by (7.20) that

$$\widetilde{A_1} \cdot \widetilde{A_2} \cdot \widetilde{A_3} \cdot \widetilde{A_4} \cdot \widetilde{A_5}' = 2k\widetilde{G}.$$

Since $\mathbf{h}_5' \in \langle \mathbf{g}_1 \rangle$ and $\mathbf{h}_5' \in \langle \mathbf{g}_2 \rangle$, exchanging the subscripts 4 and 5, it follows by the result proved above that $k$ is divisible by an odd prime. *As a conclusion, in this subcase k must be divisible by an odd prime.*

**Subcase 2.2.** $\overline{A_1} \cdot \overline{A_2} \cdot \overline{A_3} \cdot \overline{A_4} = 0$ *has a primitive solution of* $(3,3,3,3)$ *type.* First, by the proof of Lemma 7.3, we have

$$\begin{cases} \mathbf{h}_1^3 = \mathbf{h}_2^3 = \mathbf{h}_3^3 = \mathbf{h}_4^3 = \mathbf{1} \\ \mathbf{h}_1\mathbf{h}_2\mathbf{h}_3 = \mathbf{1} \\ \mathbf{h}_1\mathbf{h}_4 = \mathbf{h}_2 \\ \mathbf{h}_2\mathbf{h}_4 = \mathbf{h}_3 \\ \mathbf{h}_3\mathbf{h}_4 = \mathbf{h}_1 \end{cases} \tag{7.26}$$

and

$$\widetilde{A_1} \cdot \overline{A_2} \cdot \overline{A_3} \cdot \overline{A_4} \cdot \overline{A_5} = 0.$$

Then it can be deduced that

$$\widetilde{A_1}\,(\mathbf{1}+\mathbf{h}_1+\mathbf{h}_2\mathbf{h}_3)\,(\mathbf{1}+\mathbf{h}_2\mathbf{h}_5) = \widetilde{A_1}(\mathbf{1}+\mathbf{h}_1+\mathbf{h}_2\mathbf{h}_3)(\mathbf{h}_2+\mathbf{h}_5)$$

and, therefore, multiplying both sides by $\mathbf{h}_1$

$$\widetilde{A_1}\,(\mathbf{1}+\mathbf{h}_1+\mathbf{h}_1^2)\,(\mathbf{1}+\mathbf{h}_2\mathbf{h}_5) = \widetilde{A_1}\,(\mathbf{1}+\mathbf{h}_1+\mathbf{h}_1^2)\,(\mathbf{h}_2+\mathbf{h}_5).$$

Hence we get $\mathbf{h}_2 \in \langle \mathbf{g}_1 \rangle$ or $\mathbf{h}_5 \in \langle \mathbf{g}_1 \rangle$. Note the only solutions of $\mathbf{x}^3 = \mathbf{1}$ in $\langle \mathbf{g}_1 \rangle$ are $\mathbf{x} = \mathbf{1}$, $\mathbf{h}_1$, and $\mathbf{h}_1^2$. If $\mathbf{h}_2 \in \langle \mathbf{g}_1 \rangle$, by (7.26), then either $\mathbf{h}_2 = \mathbf{h}_1$ and hence $\mathbf{h}_4 = \mathbf{1}$, or else $\mathbf{h}_2 = \mathbf{h}_1^2$ and hence $\mathbf{h}_3 = \mathbf{1}$. Therefore, we must have $\mathbf{h}_5 \in \langle \mathbf{g}_1 \rangle$. Similarly, we also get $\mathbf{h}_5 \in \langle \mathbf{g}_2 \rangle$ and $\mathbf{h}_5 \in \langle \mathbf{g}_3 \rangle$.

In addition, it follows by $\overline{A_1} \cdot \overline{A_2} \cdot \overline{A_3} \cdot \widetilde{A_4} \cdot \overline{A_5} = 0$ that

$$\widetilde{A_4}(\mathbf{h}_1+\mathbf{h}_2+\mathbf{h}_3)(\mathbf{h}_2+\mathbf{h}_5) = \widetilde{A_4}(\mathbf{h}_1+\mathbf{h}_2+\mathbf{h}_3)(\mathbf{1}+\mathbf{h}_2\mathbf{h}_5).$$

Multiplying both sides by $\mathbf{h}_1^{-1}$, this yields

$$\widetilde{A_4}\,(\mathbf{1}+\mathbf{h}_4+\mathbf{h}_4^2)\,(\mathbf{h}_2+\mathbf{h}_5) = \widetilde{A_4}\,(\mathbf{1}+\mathbf{h}_4+\mathbf{h}_4^2)\,(\mathbf{1}+\mathbf{h}_2\mathbf{h}_5).$$

Thus we have either $\mathbf{h}_2 \in \langle \mathbf{g}_4 \rangle$ or $\mathbf{h}_5 \in \langle \mathbf{g}_4 \rangle$. If $\mathbf{h}_2 \in \langle \mathbf{g}_4 \rangle$, by (7.26), then either $\mathbf{h}_2 = \mathbf{h}_4$ and hence $\mathbf{h}_1 = \mathbf{1}$, or else $\mathbf{h}_2 = \mathbf{h}_2^2$ and hence $\mathbf{h}_3 = \mathbf{1}$. Thus we must have $\mathbf{h}_5 \in \langle \mathbf{g}_4 \rangle$. So we have $\mathbf{h}_5 \in \langle \mathbf{g}_i \rangle$ for all $i \leq 5$.

Now we write $s = |\mathbf{h}_5|$. We proceed to show that $s$ is not divisible by 3. If, on the contrary, $s = 3t$, then we have $(\mathbf{h}_5^t)^3 = \mathbf{1}$ and therefore $\mathbf{h}_5^t = \mathbf{h}_i$ or $\mathbf{h}_i^2$

for $i \le 3$. Thus two of the three elements $\mathbf{h}_1$, $\mathbf{h}_2$, and $\mathbf{h}_3$ would be equal and hence $\mathbf{h}_4 = \mathbf{1}$.

Since $\mathbf{h}_5 \in \langle \mathbf{g}_i \rangle$, we have $s \mid |\mathbf{g}_i|$ and $|\mathbf{g}_i| \mid 3q_i$. Thus we get $s \mid q_i$ for all $i \le 4$. Write $q_i = sd_i$. It is easy to see that all solutions of $\mathbf{x}^s = \mathbf{1}$ in $\langle \mathbf{g}_i \rangle$ are powers of $\mathbf{h}_5$, therefore $\mathbf{g}_i^{3d_i} \in \langle \mathbf{h}_5 \rangle$ for all $i \le 4$.

Then it follows by (7.20) that

$$\widetilde{A_1}^* \cdot \widetilde{A_2}^* \cdot \widetilde{A_3}^* \cdot \widetilde{A_4}^* \cdot \widetilde{A_5}^* = 3^4 sk\, \widetilde{G},$$

which leads to

$$s^4 \prod_{i=1}^{4} \left( \mathbf{1} + \mathbf{g}_i + \cdots + \mathbf{g}_i^{3d_i - 1} \right) \widetilde{A_5}^* = 81 sk\, \widetilde{G}.$$

Thus we have $s^3 \mid k$. *In other words, in this subcase the multiplicity $k$ is divisible by the cube of a prime other than* 3.

**Subcase 2.3.** $\overline{A_1} \cdot \overline{A_2} \cdot \overline{A_3} \cdot \overline{A_4} = 0$ *has a primitive solution of* $(2, 2, 4, 4)$ *type.* First, by the proof of Lemma 7.3, we have

$$\begin{cases} \mathbf{h}_1^2 = \mathbf{h}_2^2 = \mathbf{h}_3^4 = \mathbf{h}_4^4 = \mathbf{1} \\ \mathbf{h}_2 \mathbf{h}_3 \mathbf{h}_4 = \mathbf{1} \\ \mathbf{h}_1 \mathbf{h}_4 = \mathbf{h}_3 \\ \mathbf{h}_3 \mathbf{h}_4 = \mathbf{h}_2 \\ \mathbf{h}_1 \mathbf{h}_3 = \mathbf{h}_4 \end{cases} \tag{7.27}$$

and

$$\widetilde{A_1} \cdot \overline{A_2} \cdot \overline{A_3} \cdot \overline{A_4} \cdot \overline{A_5} = 0.$$

Then it can be deduced that

$$\widetilde{A_1} (\mathbf{h}_3 + \mathbf{h}_4)(\mathbf{h}_2 + \mathbf{h}_5) = \widetilde{A_1} (\mathbf{h}_3 + \mathbf{h}_4)(\mathbf{1} + \mathbf{h}_2 \mathbf{h}_5)$$

and, therefore, multiplying both sides by $\mathbf{h}_3^{-1}$

$$\widetilde{A_1} (\mathbf{1} + \mathbf{h}_1)(\mathbf{h}_2 + \mathbf{h}_5) = \widetilde{A_1} (\mathbf{1} + \mathbf{h}_1)(\mathbf{1} + \mathbf{h}_2 \mathbf{h}_5).$$

Hence we get $\mathbf{h}_2 \in \langle \mathbf{g}_1 \rangle$ or $\mathbf{h}_5 \in \langle \mathbf{g}_1 \rangle$. If $\mathbf{h}_2 \in \langle \mathbf{g}_1 \rangle$, then $\mathbf{h}_2 = \mathbf{h}_1$. It follows by the last four equations of (7.27) and the proof of Lemma 7.2 that $\overline{A_2} \cdot \overline{A_3} \cdot \overline{A_4} = 0$ has a primitive solution of type $(2, 2, 2)$, which contradicts our assumption. Therefore, we must have $\mathbf{h}_5 \in \langle \mathbf{g}_1 \rangle$. Similarly, we also get $\mathbf{h}_5 \in \langle \mathbf{g}_2 \rangle$.

In addition, it follows by $\overline{A_1} \cdot \overline{A_2} \cdot \widetilde{A_3} \cdot \overline{A_4} \cdot \overline{A_5} = 0$ that

$$\widetilde{A_3} (\mathbf{1} + \mathbf{h}_3)(\mathbf{1} + \mathbf{h}_1 \mathbf{h}_2 + \mathbf{h}_1 \mathbf{h}_5 + \mathbf{h}_2 \mathbf{h}_5) = \widetilde{A_3} (\mathbf{1} + \mathbf{h}_3)(\mathbf{h}_1 + \mathbf{h}_2 + \mathbf{h}_5 + \mathbf{h}_1 \mathbf{h}_2 \mathbf{h}_5).$$

Since $\mathbf{h}_1\mathbf{h}_4 \cdot \mathbf{h}_2 = \mathbf{h}_3 \cdot \mathbf{h}_3\mathbf{h}_4$ and hence $\mathbf{h}_1\mathbf{h}_2 = \mathbf{h}_3^2$, we can deduce from the above equation that

$$\widetilde{A}_3(\mathbf{1}+\mathbf{h}_3)\left(\mathbf{1}+\mathbf{h}_3^2\right)(\mathbf{1}+\mathbf{h}_2\mathbf{h}_5) = \widetilde{A}_3(\mathbf{1}+\mathbf{h}_3)\left(\mathbf{1}+\mathbf{h}_3^2\right)(\mathbf{h}_1+\mathbf{h}_5).$$

Then we have either $\mathbf{h}_1 \in \langle \mathbf{g}_3 \rangle$ or $\mathbf{h}_5 \in \langle \mathbf{g}_3 \rangle$. If $\mathbf{h}_1 \in \langle \mathbf{g}_3 \rangle$, since $\mathbf{h}_1\mathbf{h}_2 = \mathbf{h}_3^2$, then we also get $\mathbf{h}_2 \in \langle \mathbf{h}_3 \rangle$. Since $\mathbf{h}_1^2 = \mathbf{h}_2^2 = \mathbf{1}$, we would find $\mathbf{h}_1 = \mathbf{h}_2$, which is impossible. Thus we must have $\mathbf{h}_5 \in \langle \mathbf{g}_3 \rangle$ and similarly $\mathbf{h}_5 \in \langle \mathbf{g}_4 \rangle$. So we have $\mathbf{h}_5 \in \langle \mathbf{g}_i \rangle$ for all $i \le 5$.

Now we write $s = |\mathbf{h}_5|$. As in the other subcases, it can be deduced that $s$ is odd, $s \mid |\mathbf{g}_i|$, and $|\mathbf{g}_i| \mid q_i|\mathbf{h}_i|$. Since $|\mathbf{h}_i| = 2$ or $4$ for $i \le 4$, we get $s \mid q_i$. Write $q_i = sd_i$. Since all solutions of $\mathbf{x}^s = \mathbf{1}$ in $\langle \mathbf{g}_i \rangle$ are powers of $\mathbf{h}_5$, we get $\mathbf{g}_i^{|\mathbf{h}_i|d_i} \in \langle \mathbf{h}_5 \rangle$ for all $i \le 4$.

Then it follows by (7.20) that

$$\widetilde{A}_1^{\,*} \cdot \widetilde{A}_2^{\,*} \cdot \widetilde{A}_3^{\,*} \cdot \widetilde{A}_4^{\,*} \cdot \widetilde{A}_5^{\,*} = 2^2 \cdot 4^2 sk\, \widetilde{G},$$

which leads to

$$s^4 \prod_{i=1}^{4} \left(\mathbf{1}+\mathbf{g}_i+\cdots+\mathbf{g}_i^{|\mathbf{h}_i|d_i-1}\right) \widetilde{A}_5^{\,*} = 64sk\,\widetilde{G}.$$

Thus we have $s^3 \mid k$. *In other words, in this subcase the multiplicity $k$ is divisible by the cube of an odd prime.*

As a conclusion of these three subcases, the necessary part of Case 2 is proved.

To prove the sufficient part, we have two constructions corresponding to the first two subcases.

**Construction 1.** Assume that $p$ is an odd prime and $q$ is a positive integer. Let us take $G = O_{2p} \otimes O_4$ and write $\mathbf{e}_1 = (e^{\pi i/p}, 1)$ and $\mathbf{e}_2 = (1, e^{\pi i/2})$. Clearly, $|\mathbf{e}_1| = 2p$ and $|\mathbf{e}_2| = 4$. We define

$$\begin{aligned}
\widetilde{A}_1 &= \mathbf{1}+\mathbf{e}_1+\cdots+\mathbf{e}_1^{pq-1} \\
&= \left(\mathbf{1}+\mathbf{e}_1+\cdots+\mathbf{e}_1^{p-1}\right)\left(\mathbf{1}+\mathbf{e}_1^{p}+\cdots+\mathbf{e}_1^{p(q-1)}\right), \\
\widetilde{A}_2 &= \mathbf{1}+\mathbf{e}_1\mathbf{e}_2^2+\cdots+\left(\mathbf{e}_1\mathbf{e}_2^2\right)^{p-1}, \\
\widetilde{A}_3 &= \mathbf{1}+\mathbf{e}_2, \\
\widetilde{A}_4 &= \mathbf{1}+\mathbf{e}_1^{p}\mathbf{e}_2,
\end{aligned}$$

and

$$\widetilde{A}_5 = \mathbf{1}+\mathbf{e}_1^2\mathbf{e}_2^2.$$

Then it can be verified that

$$\begin{cases} \widetilde{A}_1 \cdot \widetilde{A}_2 \cdot \widetilde{A}_3 \cdot \widetilde{A}_4 \cdot \widetilde{A}_5 = pq\,\widetilde{G} \\ \mathbf{h}_i \neq \mathbf{1}, \quad i = 1, 2, \ldots, 5. \end{cases}$$

**Construction 2.** Assume that $q$ is a positive integer. Let us take $G = O_{18} \otimes O_3$ and write $\mathbf{e}_1 = (e^{\pi i/9}, 1)$ and $\mathbf{e}_2 = (1, e^{2\pi i/3})$. Clearly, $|\mathbf{e}_1| = 18$ and $|\mathbf{e}_2| = 3$. We define

$$\begin{aligned} \widetilde{A}_1 &= \mathbf{1} + \mathbf{e}_1^3\mathbf{e}_2^2 + \cdots + (\mathbf{e}_1^3\mathbf{e}_2^2)^{2q-1} \\ &= (\mathbf{1} + \mathbf{e}_1^3\mathbf{e}_2^2)(\mathbf{1} + (\mathbf{e}_1^3\mathbf{e}_2^2)^2 + \cdots + (\mathbf{e}_1^3\mathbf{e}_2^2)^{2(q-1)}), \\ \widetilde{A}_2 &= \mathbf{1} + \mathbf{e}_1^{15}\mathbf{e}_2^2, \\ \widetilde{A}_3 &= \mathbf{1} + \mathbf{e}_1^9\mathbf{e}_2^2, \\ \widetilde{A}_4 &= \mathbf{1} + \mathbf{e}_1 + \cdots + \mathbf{e}_1^5, \end{aligned}$$

and

$$\widetilde{A}_5 = \mathbf{1} + \mathbf{e}_1\mathbf{e}_2 + \cdots + (\mathbf{e}_1\mathbf{e}_2)^8.$$

Then it can be verified that

$$\begin{cases} \widetilde{A}_1 \cdot \widetilde{A}_2 \cdot \widetilde{A}_3 \cdot \widetilde{A}_4 \cdot \widetilde{A}_5 = 8q\,\widetilde{G} \\ \mathbf{h}_i \neq \mathbf{1}, \quad i = 1, 2, \ldots, 5. \end{cases}$$

Let $Z^*$ denote the set of the positive integers and let $Z^\star$ denote the set of the positive integers, which is divisible by eight or by an odd prime. It is easy to see that

$$\{1, 2, 4\} = Z^* \setminus Z^\star.$$

Thus the sufficient part follows by the two constructions. The second case is proved. □

**Proof of Theorem 7.3, Case 3.** The necessary part has been proved in Chapter 6. To show the sufficient part, we have the following construction for $n = 6$. The higher-dimensional case follows inductively. Assume that $q$ is a positive integer. Let us take $G = O_6 \otimes O_6 \otimes O_3$ and write $\mathbf{e}_1 = (e^{\pi i/3}, 1, 1)$, $\mathbf{e}_2 = (1, e^{\pi i/3}, 1)$, and $\mathbf{e}_3 = (1, 1, e^{2\pi i/3})$. Clearly, $|\mathbf{e}_1| = 6$, $|\mathbf{e}_2| = 6$, and $|\mathbf{e}_3| = 3$. Then we define

$$\begin{aligned} \widetilde{A}_1 &= \mathbf{1} + \mathbf{e}_1 + \cdots + \mathbf{e}_1^{3q-1} \\ &= (\mathbf{1} + \mathbf{e}_1 + \mathbf{e}_1^2)\left(\mathbf{1} + \mathbf{e}_1^3 + \cdots + \mathbf{e}_1^{3(q-1)}\right), \\ \widetilde{A}_2 &= \mathbf{1} + \mathbf{e}_2 + \mathbf{e}_2^2, \\ \widetilde{A}_3 &= \mathbf{1} + \mathbf{e}_1^2\mathbf{e}_2^3, \\ \widetilde{A}_4 &= \mathbf{1} + \mathbf{e}_1^3\mathbf{e}_2^2, \\ \widetilde{A}_5 &= \mathbf{1} + \mathbf{e}_1\mathbf{e}_2^3\mathbf{e}_3 + (\mathbf{e}_1\mathbf{e}_2^3\mathbf{e}_3)^2, \end{aligned}$$

and

$$\widetilde{A}_6 = \mathbf{1} + \mathbf{e}_1^4 \mathbf{e}_2^3 \mathbf{e}_3.$$

It can be verified that

$$\begin{cases} \widetilde{A}_1 \cdot \widetilde{A}_2 \cdot \widetilde{A}_3 \cdot \widetilde{A}_4 \cdot \widetilde{A}_5 \cdot \widetilde{A}_6 = 2q\,\widetilde{G} \\ \mathbf{h}_i \neq \mathbf{1}, \quad i = 1, 2, \ldots, 6, \end{cases}$$

which means that Furtwängler's conjecture is false when $n = 6$ and $k$ is even.

By Case 2, it follows that the conjecture is also false when $n = 6$ and $k \neq 1$ is odd. Hence the third case is proved. □

# 8

# Keller's conjecture

## 8.1 Keller's conjecture

In 1930, ten years before Hajós' proof, O. Keller generalized Minkowski's conjecture from lattice tilings to translative tilings and made the following conjecture.

**Keller's conjecture.** *Every translative tiling $I^n + X$ of $E^n$ has a twin.*

As was shown at the beginning of Chapter 6, the two- and three-dimensional cases of this conjecture can be proved by routine arguments. In 1940, Perron (1940a) extended this result to $n \leq 6$ through elementary but complicated arguments. In fact, Keller himself did claim a proof for the $n \leq 6$ cases in 1937. However, his sketch is hardly accepted as a proof.

In 1949, based on an ingenious observation, G. Hajós was able to formulate Keller's conjecture into the following algebraic version[8].

*Let $G$ be an abelian group generated by $\mathbf{g}_1, \mathbf{g}_2, \ldots, \mathbf{g}_n$ with $|\mathbf{g}_i| = 2q_i$. If $G = H \cdot A_1 \cdots A_n$ is a factorization, where $|H| = 2^n$ and $A_i = \{\mathbf{1}, \mathbf{g}_i, \ldots, \mathbf{g}_i^{q_i - 1}\}$, then*

$$\left\{\mathbf{h}_i \mathbf{h}_j^{-1} : \mathbf{h}_i, \mathbf{h}_j \in H\right\} \bigcap \left\{\mathbf{g}_i^{q_i} : i \leq n\right\} \neq \emptyset.$$

In 1980s, while no essential progress was made in the positive direction, S. Szabó and K. Corrádi started to consider this problem in the negative direction. First, Szabó (1986) proved that, to search for a counterexample for Keller's conjecture, it is sufficient to restrict $q_i = 2$ for all $i \leq n$ in the algebraic version. Then Corrádi and Szabó (1990b) obtained a criterion in graph theory for such a counterexample, if it does exist.

[8] In this chapter, we say $G = A_1 A_2 \cdots A_n$ (or $G = A_1 + A_2 + \cdots + A_n$) is a factorization if every element $\mathbf{g} \in G$ can be uniquely expressed as $\mathbf{g} = \mathbf{a}_1 \mathbf{a}_2 \cdots \mathbf{a}_n$ (or $\mathbf{g} = \mathbf{a}_1 + \mathbf{a}_2 + \cdots + \mathbf{a}_n$), where $\mathbf{a}_i \in A_i$.

Consider the set $\{\mathbf{z} = (z_1, z_2, \ldots, z_n) : z_i \in Z_4\}$, where $Z_4 = \{0, 1, 2, 3\}$, and define a graph $\Gamma_n$ on it in the following way. Two vertices $\mathbf{z}$ and $\mathbf{z}'$ are adjacent if $z_i - z'_i \equiv 2 \pmod 4$ for some $i$ and $z_j \neq z'_j$ for some $j$, where $j \neq i$. It is easy to see that $\Gamma_n$ has $4^n$ vertices. A subgraph of $\Gamma_n$ is called a *clique* if any two of its vertices are adjacent. The *size* of a clique is the cardinality of its vertices. For convenience, let $\rho_n$ denote the maximal size of a clique in $\Gamma_n$. With these definitions, *Corrádi and Szabó's criterion* can be stated as follows.

*If $\rho_n = 2^n$, then there exists a counterexample for Keller's conjecture in $E^n$.*

Based on this criterion, in 1992 J.C. Lagarias and P. Shor were able to construct such an example for $n \geq 10$ and therefore got the first counterexample for Keller's conjecture. In 2002, the dimensions were reduced to $n \geq 8$ by J. Mackey. So far, Keller's conjecture is open only for $n = 7$. Therefore, the known results about Keller's conjecture can be concluded as follows.

*When $n \leq 6$, Keller's conjecture is true; whenever $n \geq 8$ it is false.*

## 8.2 A theorem of Keller and Perron

In 1937, O.H. Keller sketched a proof for his conjecture for $n \leq 6$. In 1940, O. Perron gave a complete proof for this result.

**Theorem 8.1 (Keller, 1937; Perron, 1940a).** *When $n \leq 6$, every translative tiling $I^n + X$ of $E^n$ has a twin.*

As we can imagine, this theorem was proved by complicated case-by-case arguments. We will not reproduce the whole proof here. Instead, we will introduce some general results, which did play important roles in the proof. Then we will demonstrate how to apply them to deduce the two- and three-dimensional cases of the theorem.

For convenience, for a discrete set $X$ in $E^n$, two fixed integers $i$ and $u$ with $0 \leq i \leq n$, and a fixed fractional $v$, we define

$$X_i(u, v) = \Big\{\mathbf{x} \in X : \ [x_i] = u \text{ and } \{x_i\} = v\Big\},$$

where $[x]$ and $\{x\}$ denote the integer part and the fractional part of $x$, respectively.

**Lemma 8.1 (Keller, 1937).** *If $I^n + X$ is a tiling of $E^n$ and $v$ is a fixed fractional number, then*

$$I^n + X_i(u, v) = \Big(I^n + X_i(u', v)\Big) + (u - u')\mathbf{e}_i$$

*holds for any set of integers $u$, $u'$, and $i$ with $1 \leq i \leq n$.*

**Proof.** It is sufficient to show the $i = 1$ case. Let $H^t$ denote the hyperplane $\{\mathbf{x} \in E^n : x_n = t\}$, and let $\overline{H^t}$ denote the halfspace bounded by $H^t$ and containing $(t+1)\mathbf{e}_n$. Then the system

$$\{H^t \cap (I^n + \mathbf{x}) : (\text{int}(I^n) + \mathbf{x}) \cap \overline{H^t} \neq \emptyset\}$$

is a tiling of $H^t$. For convenience, we write this tiling as $I^{n-1} + X^t$.

Clearly the lemma is true when $n = 1$. Assume that it is true in $(n-1)$ dimensions, for all $u, u' \in Z$ we have

$$I^{n-1} + X_1^t(u, v) = \left(I^{n-1} + X_1^t(u', v)\right) + (u - u')\mathbf{e}_1.$$

On the other hand, it is easy to see that

$$I^{n-1} + X_1^t(u, v) = \Big\{(I^n + \mathbf{x}) \cap H^t : \mathbf{x} \in X_1(u, v), \\ (\text{int}(I^n) + \mathbf{x}) \cap \overline{H^t} \neq \emptyset\Big\}.$$

Thus we get

$$\begin{aligned} I^n + X_1(u, v) &= \bigcup_{-\infty < t < \infty} \left(I^{n-1} + X_1^t(u, v)\right) \\ &= \bigcup_{-\infty < t < \infty} \left[\left(I^{n-1} + X_1^t(u', v)\right) + (u - u')\mathbf{e}_1\right] \\ &= \left(I^n + X_1(u', v)\right) + (u - u')\mathbf{e}_1. \end{aligned}$$

The lemma is proved. □

Geometrically speaking, the union of the cubes $I^n + \mathbf{x}$, $\mathbf{x} \in X$, with fixed $\{x_i\}$ forms a cylinder of infinite length in the direction of $\mathbf{e}_i$. Therefore, we have the following corollary.

**Corollary 8.1 (Szabó, 1986).** *If $I^n + X$ is a tiling of $E^n$, then*

$$\bigcup_u \left(I^n + X_i(u, v)\right) = \bigcup_u \left(I^n + X_i(u, v) + w\mathbf{e}_i\right)$$

*holds for any fixed $i$, $v$, and $w$.*

**Lemma 8.2 (Keller, 1930).** *Let $I^n + X$ be a tiling of $E^n$ and let $\mathbf{x}$ and $\mathbf{y}$ be two points of $X$. If*

$$|x_i - y_i| < 1$$

*holds for all $i \le n - 1$, then $x_n - y_n$ is an integer.*

**Proof.** Without loss of generality, we write

$$x_n - y_n = z + \alpha,$$

where $z$ is an positive integer and $\alpha$ is a nonnegative fractional. Now we proceed to show $\alpha = 0$.

If $z = 1$ and $0 < \alpha < 1$, then there is a point $\mathbf{x}' \in X \setminus \{\mathbf{x}, \mathbf{y}\}$, a small positive number $\epsilon$, and a point $\mathbf{p}$ which is very close to $\frac{1}{2}(\mathbf{x}+\mathbf{y})$ such that

$$\epsilon B^n + \mathbf{p} \subset I^n + \mathbf{x}'.$$

For convenience, we write

$$C = \left\{\epsilon B^n + \mathbf{p} + t\mathbf{e}_n : \ -\infty < t < \infty\right\} \bigcap \left(I^n + \mathbf{x}'\right).$$

Since the height of $C$ is 1 and $\alpha < 1$, the cylinder $C$ will intersect the interior of either $I^n + \mathbf{x}$ or $I^n + \mathbf{y}$, which contradicts the assumption that $I^n + X$ is a tiling. Therefore, we must have $\alpha = 0$.

Assume that the assertion is true when $z \leq m - 1$, we proceed to prove the $z = m$ case by induction. First, we claim that there is a point $\mathbf{y}' \in X$ satisfying

$$\begin{cases} y_i \leq y_i' < y_i + 1 & \text{if } y_i \leq x_i \text{ and } i \leq n-1, \\ y_i - 1 < y_i' \leq y_i & \text{if } x_i < y_i \text{ and } i \leq n-1, \\ y_n + m - 1 \leq y_n' < y_n + m. \end{cases}$$

Otherwise, we can find a point which is not covered by $I^n + X$ and therefore $I^n + X$ is not a tiling. Then we get

$$\begin{cases} |x_i - y_i'| < 1 & \text{if } i \leq n-1, \\ \alpha < x_n - y_n' \leq 1 + \alpha. \end{cases}$$

It follows by the $z = 1$ case that $x_n - y_n' = 1$ and, consequently

$$\begin{cases} |y_i' - y_i| < 1 & \text{if } i \leq n-1, \\ y_n' - y_n = m - 1 + \alpha. \end{cases}$$

The lemma follows by the induction. □

**Lemma 8.3 (Keller, 1930).** *Let $I^n + X$ be a tiling of $E^n$. For any two points $\mathbf{x}$ and $\mathbf{y}$ of $X$, the vector $\mathbf{x} - \mathbf{y}$ has at least one nonzero integer coordinate.*

**Proof.** If $|x_i - y_i| \geq 1$ holds for only one index, then it is nothing else but Lemma 8.2. Assume that the statement is true when

$$|x_i - y_i| \geq 1 \tag{8.1}$$

holds for $k - 1$ indices, we proceed to show that it is also true when (8.1) holds for $k$ indices.

Without loss of generality, we assume

$$\begin{cases} |x_i - y_i| \geq 1 & \text{if } i \leq k, \\ |x_i - y_i| < 1 & \text{if } i > k. \end{cases}$$

If $|x_1 - y_1|$ is an integer, then there is nothing to prove. If $|x_1 - y_1|$ is not an integer, then, by Lemma 8.1, we can get a new tiling $I^n + X'$ such that both $\mathbf{x}' = (x_1', x_2', \ldots, x_n') = (x_1 + [y_1 - x_1], x_2, \ldots, x_n)$ and $\mathbf{y}$ belong to $X'$. Now we get

$$\begin{cases} |x_i' - y_i| \geq 1 & \text{if } 2 \leq i \leq k, \\ |x_i' - y_i| < 1 & \text{otherwise.} \end{cases}$$

By the inductive assumption, the lemma follows. □

**Proof of Theorem 8.1, the two-dimensional case.** Let $I^2 + X$ be a tiling of $E^2$. We may assume that $X$ has three points $\mathbf{x}_0 = (0, 0)$, $\mathbf{x}_1 = (1 + \alpha_{11}, \alpha_{12})$, and $\mathbf{x}_2 = (\alpha_{21}, 1 + \alpha_{22})$, where $0 \leq \alpha_{ij} < 1$. For example, if no such $\mathbf{x}_1$ can be found, then we can get a suitable $\epsilon > 0$ such that

$$(1.5 - \epsilon, 0.5 - \epsilon) \notin I^2 + X,$$

which contradicts the fact that $I^2 + X$ is a tiling of $E^2$. In fact, for any $\alpha$ and $\beta$, the square $\{\mathbf{x} : \ \alpha \leq x_1 < \alpha + 1; \ \beta \leq x_2 < \beta + 1\}$ contains a point of $X$.

Applying Lemma 8.3 to $\{\mathbf{x}_0, \mathbf{x}_1\}$ and $\{\mathbf{x}_0, \mathbf{x}_2\}$, we get $\alpha_{11} = \alpha_{22} = 0$. Then apply it to $\{\mathbf{x}_1, \mathbf{x}_2\}$, we get either $\alpha_{12} = 0$ or $\alpha_{21} = 0$. If $\alpha_{12} = 0$, then $I^2 + \mathbf{x}_0$ and $I^2 + \mathbf{x}_1$ form a twin. Otherwise $I^2 + \mathbf{x}_0$ and $I^2 + \mathbf{x}_2$ will form a twin. The two-dimensional case is proved. □

**Proof of Theorem 8.1, the three-dimensional case.** Let $I^3 + X$ be a tiling of $E^3$. We may assume that $X$ has five points $\mathbf{x}_0 = (0, 0, 0)$, $\mathbf{x}_1 = (1 + \beta_{11}, \beta_{12}, \beta_{13})$, $\mathbf{x}_2 = (\beta_{21}, 1 + \beta_{22}, \beta_{23})$, $\mathbf{x}_3 = (\beta_{31}, \beta_{32}, 1 + \beta_{33})$, and $\mathbf{x}_4 = (1 + \beta_{41}, 1 + \beta_{42}, \beta_{43})$, where $0 \leq \beta_{ij} < 1$.

Applying Lemma 8.3 to $\{\mathbf{x}_0, \mathbf{x}_1\}$, $\{\mathbf{x}_0, \mathbf{x}_2\}$, and $\{\mathbf{x}_0, \mathbf{x}_3\}$, we get $\beta_{11} = \beta_{22} = \beta_{33} = 0$, to $\{\mathbf{x}_1, \mathbf{x}_2\}$, $\{\mathbf{x}_1, \mathbf{x}_3\}$, and $\{\mathbf{x}_2, \mathbf{x}_3\}$, we get either $\beta_{ij} = 0$ or $\beta_{ji} = 0$ for all $\{i, j\} = \{1, 2\}$, $\{1, 3\}$, and $\{2, 3\}$. Without loss of generality, we assume $\beta_{21} = 0$. Now we deal with the following cases.

**Case 1.** $\beta_{23} = 0$, then $I^3 + \mathbf{x}_0$ and $I^3 + \mathbf{x}_2$ form a twin.
**Case 2.** $\beta_{32} = \beta_{31} = 0$, then $I^3 + \mathbf{x}_0$ and $I^3 + \mathbf{x}_3$ form a twin.
**Case 3.** $\beta_{32} = \beta_{13} = \beta_{12} = 0$, then $I^3 + \mathbf{x}_0$ and $I^3 + \mathbf{x}_1$ form a twin.
**Case 4.** $\beta_{12} \neq 0$, $\beta_{23} \neq 0$, *and* $\beta_{31} \neq 0$. Applying Lemma 8.3 to $\{\mathbf{x}_1, \mathbf{x}_4\}$, $\{\mathbf{x}_2, \mathbf{x}_4\}$, and $\{\mathbf{x}_3, \mathbf{x}_4\}$, we get $\beta_{42} = \beta_{12}$, $\beta_{41} = 0$, and $\beta_{43} = 0$. Then $I^3 + \mathbf{x}_1$ and $I^3 + \mathbf{x}_4$ form a twin.

The three-dimensional case is proved. □

## 8.3 Corrádi and Szabó's criterion

If there are counterexamples for Keller's conjecture, then there should be quite regular ones. Based on this belief, Perron (1940a) made the first attempt in this direction. Then, Hajós (1949), Szabó (1986), and Corrádi and Szabó (1990b) made essential contributions and finally got a combinatorial criterion.

**Corrádi and Szabó's criterion.** *If there exists a counterexample for Keller's conjecture, then $\rho_m = 2^m$ holds for some m. On the other hand, if $\rho_n = 2^n$ holds, then there is a counterexample for Keller's conjecture in $E^n$.*

To deduce this criterion, we need several lemmas.

**Lemma 8.4 (Szabó, 1986).** *If there exists a counterexample for Keller's conjecture in $E^n$, then there is such an example $I^n + Y$ that $Y$ is a periodic set with periods $2\boldsymbol{e}_1, 2\boldsymbol{e}_2, \ldots, 2\boldsymbol{e}_n$.*

**Proof.** Assume that $I^n + X$ is a counterexample for Keller's conjecture. We define

$$X_1 = \bigcup_{u\in Z}\left(\bigcup_{v}(X_1(0, v)\bigcup X_1(1, v)) + 2u\mathbf{e}_1\right)$$

and proceed to show that $I^n + X_1$ is also a counterexample.

First, by Lemma 8.1, we have

$$\begin{aligned}
E^n &= I^n + X = I^n + \bigcup_{u,v} X_1(u, v) \\
&= \left(I^n + \bigcup_{u,v} X_1(2u, v)\right)\bigcup\left(I^n + \bigcup_{u,v} X_1(2u+1, v)\right) \\
&= \left[\bigcup_{u}\left(I^n + \bigcup_{v} X_1(0, v) + 2u\mathbf{e}_1\right)\right] \\
&\quad \bigcup\left[\bigcup_{u}\left(I^n + \bigcup_{v} X_1(1, v) + 2u\mathbf{e}_1\right)\right] \\
&= \bigcup_{u}\left[I^n + \bigcup_{v}(X_1(0, v)\bigcup X_1(1, v)) + 2u\mathbf{e}_1\right] \\
&= I^n + \bigcup_{u}\left[\bigcup_{v}(X_1(0, v)\bigcup X_1(1, v)) + 2u\mathbf{e}_1\right] \\
&= I^n + X_1.
\end{aligned}$$

Second, we show that $I^n + X_1$ is a tiling. If, on the contrary

$$\left(\text{int}(I^n) + \mathbf{x}\right)\bigcap\left(\text{int}(I^n) + \mathbf{x}'\right) \neq \emptyset \tag{8.2}$$

holds for two distinct points $\mathbf{x}$ and $\mathbf{x}'$ of $X_1$, then we have

$$|x_i - x_i'| < 1$$

for all $i \leq n$. Assume that $\mathbf{x} \in X_1(k, v) + 2u\mathbf{e}_1$ and $\mathbf{x}' \in X_1(k', v') + 2u'\mathbf{e}_1$, where $k, k' \in \{0, 1\}$, then

$$\mathbf{x} - \mathbf{x}' \in X_1(k, v) - X_1(k', v') + 2(u - u')\mathbf{e}_1. \tag{8.3}$$

If $v \neq v'$, then we get

$$\left(\text{int}(I^n) + X_1(2u + k, v)\right) \cap \left(\text{int}(I^n) + X_1(2u' + k', v')\right) \neq \emptyset,$$

which contradicts the assumption that $I^n + X$ is a tiling. Hence we have $v = v'$ and

$$|x_1 - x_1'| = |k - k' + 2(u - u')| < 1,$$

from which one can easily deduce $k = k'$ and $u = u'$. Then, by (8.3), we get

$$\mathbf{x} - \mathbf{x}' \in X_1(k, v) - X_1(k, v) \subset X - X,$$

which violates the assumption that $I^n + X$ is a tiling. Therefore, $I^n + X_1$ is a tiling as well.

Third, we show that $I^n + X_1$ has no twin. If, on the contrary, it has a twin $I^n + \mathbf{x}$ and $I^n + \mathbf{x}'$, then we have $\mathbf{x} \in X_1(k, v) + 2u\mathbf{e}_1$ and $\mathbf{x}' \in X_1(k', v') + 2u'\mathbf{e}_1$ for some $k, k' \in \{0, 1\}$ and $\mathbf{x} - \mathbf{x}' = \mathbf{e}_j$ for some $j \leq n$. Therefore

$$\mathbf{x} - \mathbf{x}' = \mathbf{e}_j \in X_1(k, v) - X_1(k', v') + 2(u - u')\mathbf{e}_1. \tag{8.4}$$

If $j \neq 1$, by looking at the first coordinate, we get

$$0 = (k - k') + v - v' + 2(u - u'),$$

from which we can deduce $u = u'$, $v = v'$, and thus

$$\mathbf{e}_j \in X_1(k, v) - X_1(k', v) \subset X - X,$$

which contradicts the assumption that $I^n + X$ has no twin. If $j = 1$, we get

$$1 = (k - k') + (v - v') + 2(u - u')$$

and hence $v = v'$ and either $u = u'$ or $u = u' + 1$. Then, by (8.4), we get either

$$\mathbf{e}_1 \in X_1(k, v) - X_1(k', v) \subset X - X$$

or

$$-\mathbf{e}_1 \in X_1(k, v) - X_1(k', v) \subset X - X,$$

where both contradict the assumption.

As a conclusion, $I^n + X_1$ is a counterexample for Keller's conjecture and $X_1$ is a periodic set with period $2\mathbf{e}_1$. Repeating this process for all the other coordinates, the lemma follows. □

**Lemma 8.5 (Szabó, 1986).** *If there exists a counterexample $I^n + X$ for Keller's conjecture in $E^n$, then there is such an example $I^n + X^*$ that $X^*$ is a periodic set with periods $2\mathbf{e}_1, 2\mathbf{e}_2, \ldots, 2\mathbf{e}_n$ and every vector of $X^*$ has rational coordinates whose denominators are powers of 2.*

**Proof.** By Lemma 8.4, we may assume that $X$ is a periodic set with periods $2\mathbf{e}_1, 2\mathbf{e}_2, \ldots, 2\mathbf{e}_n$. Writing

$$\Lambda = \left\{ \sum 2z_i \mathbf{e}_i : \ z_i \in Z \right\}$$

and

$$Y = \left\{ \mathbf{x} \in X : \ 0 \leq x_i < 2 \right\},$$

it is easy to see that

$$X = Y + \Lambda = \bigcup_{u,v} X_1(u, v) \tag{8.5}$$

and card$\{Y\}$ is finite. Therefore, $X_1(u, v)$ has only a finite number of different indices $v$.

For these indices $v$, we define corresponding $\overline{v}$ such that $v + \overline{v}$ are distinct rational numbers satisfying $0 \leq v + \overline{v} < 1$ and their denominators are powers of 2. Then we define

$$X_1 = \bigcup_{v} \left( \bigcup_{u} X_1(u, v) + \overline{v}\mathbf{e}_1 \right)$$

and proceed to show that $I^n + X_1$ is also a counterexample for Keller's conjecture.

First, by Corollary 8.1, we have

$$\begin{aligned} E^n = I^n + X &= I^n + \bigcup_{u,v} X_1(u, v) \\ &= \bigcup_{v} \left( I^n + \bigcup_{u} X_1(u, v) \right) \\ &= \bigcup_{v} \left( I^n + \bigcup_{u} X_1(u, v) + \overline{v}\mathbf{e}_1 \right) \\ &= I^n + \bigcup_{v} \bigcup_{u} \left( X_1(u, v) + \overline{v}\mathbf{e}_1 \right) \\ &= I^n + X_1. \end{aligned}$$

Second, we claim that $I^n + X_1$ is a tiling. If, on the contrary

$$\left(\text{int}(I^n) + \mathbf{x}\right) \bigcap \left(\text{int}(I^n) + \mathbf{x}'\right) \neq \emptyset$$

holds for two distinct points $\mathbf{x}$ and $\mathbf{x}'$ of $X_1$, we proceed to get a contradiction. It is easy to see that

$$\text{int}\left(I^n + \bigcup_u X_1(u, v) + \overline{v}\mathbf{e}_1\right) \bigcap \text{int}\left(I^n + \bigcup_u X_1(u, w) + \overline{w}\mathbf{e}_1\right) = \emptyset$$

whenever $v \neq w$. Hence, if

$$\mathbf{x} \in X_1(u, v) + \overline{v}\mathbf{e}_1$$

and

$$\mathbf{x}' \in X_1(u', v') + \overline{v'}\mathbf{e}_1,$$

then we have $v = v'$ and thus

$$|x_1 - x_1'| = |u - u'| < 1.$$

Consequently, we get $u = u'$ and by (8.5)

$$\mathbf{x} - \mathbf{x}' \in X_1(u, v) - X_1(u, v) \subset X - X,$$

which contradicts the assumption that $I^n + X$ is a tiling. Thus $I^n + X_1$ is also a tiling.

Third, we claim that $I^n + X_1$ has no twin. If $\mathbf{x} \in X_1(u, v) + \overline{v}\mathbf{e}_1$, $\mathbf{x}' \in X_1(u', v') + \overline{v'}\mathbf{e}_1$, and $\mathbf{x} - \mathbf{x}' = \mathbf{e}_j$, by looking at the first coordinate, we get that $(u - u') + (v + \overline{v}) - (v' + \overline{v'})$ is either 0 or 1. Then since by assumption $v + \overline{v}$ and $v' + \overline{v'}$ are distinct fractional numbers if $v$ and $v'$ are different, it follows that $v$ and $v'$ must be identical. Therefore, by (8.5), we get

$$\mathbf{x} - \mathbf{x}' = \mathbf{e}_j \in \bigcup_u X_1(u, v) - \bigcup_u X_1(u, v) \subset X - X,$$

which contradicts the assumption that $I^n + X$ has no twin.

Repeating this process to the other coordinates the lemma follows. □

In contrast to Chapters 6 and 7, in the rest of this chapter we prefer the additive expression for an abelian group rather than the multiplicative form.

**Lemma 8.6 (Hajós, 1949; Szabó, 1986).** *If there does exist a counterexample for Keller's conjecture, then there is a finite abelian group G with a factorization*

$$G = H + [\mathbf{g}_1]_2 + \cdots + [\mathbf{g}_m]_2$$

*such that* $2\mathbf{g}_i \notin H - H$ *holds for all i.*

**Proof.** Assume that $I^n + X$ is a counterexample for Keller's conjecture. By Lemma 8.5, we may assume that $X$ is a periodic set with periods $2\mathbf{e}_1$, $2\mathbf{e}_2, \ldots, 2\mathbf{e}_n$ and every point of $X$ has rational coordinates whose denominators are powers of 2. We denote the largest denominator by $q$ and define

$$\Lambda = \left\{ \sum 2z_i \mathbf{e}_i : \ z_i \in Z \right\},$$
$$H = \left\{ \mathbf{x} \in X : \ 0 \le x_i < 2 \right\},$$

and

$$\Lambda^* = \left\{ \tfrac{1}{q} \sum z_i \mathbf{e}_i : \ z_i \in Z \right\}.$$

Since $I^n + X$ is a tiling and $X \subset \Lambda^*$, we have

$$\Lambda^* = \left[\tfrac{1}{q}\mathbf{e}_1\right]_q + \cdots + \left[\tfrac{1}{q}\mathbf{e}_n\right]_q + X$$
$$= H + \left[\tfrac{1}{q}\mathbf{e}_1\right]_q + \cdots + \left[\tfrac{1}{q}\mathbf{e}_n\right]_q + \Lambda.$$

Since both $\Lambda$ and $\Lambda^*$ are abelian groups, and $\Lambda \subset \Lambda^*$, we get a quotient group

$$G = \Lambda^*/\Lambda = H + [\mathbf{g}_1]_q + \cdots + [\mathbf{g}_n]_q, \tag{8.6}$$

where $\mathbf{g}_i$ denotes the coset corresponding to $\frac{1}{q}\mathbf{e}_i$.

First we claim that $q\mathbf{g}_i \notin H - H$ for each $i$. Otherwise, we would have

$$\mathbf{e}_i + \mathbf{u}_0 = \mathbf{h}_1 + \mathbf{u}_1 - \mathbf{h}_2 - \mathbf{u}_2,$$

where $\mathbf{u}_j \in \Lambda$ and $\mathbf{h}_j \in H$, and hence

$$\mathbf{e}_i = \mathbf{x}_1 - \mathbf{x}_2$$

for two suitable points $\mathbf{x}_1$ and $\mathbf{x}_2$ of $X$, which contradicts the assumption that $I^n + X$ has no twin.

Now we factorize $[\mathbf{g}_i]_q$. If $q = rs$, then we can replace $[\mathbf{g}_i]_q$ by $[\mathbf{g}_i]_r + [r\mathbf{g}_i]_s$ in (8.6). To see this, we only need to show $r\mathbf{g}_i \notin H - H$. If $r\mathbf{g}_i = \mathbf{h}_1 - \mathbf{h}_2$, then $\mathbf{h}_1$ can be expressed in two ways $\mathbf{h}_1 = \mathbf{h}_1$ and $\mathbf{h}_1 = \mathbf{h}_2 + r\mathbf{g}_i$, which contradicts the assumption. Thus $[\mathbf{g}_i]_{2^l}$ can be replaced by $[\mathbf{g}_i]_2 + [2\mathbf{g}_i]_2 + \cdots + [2^{l-1}\mathbf{g}_i]_2$. The lemma is proved. □

Next let us introduce a basic result in group theory, which will be useful in the further discussion.

**Lemma 8.7 (Stein, 1972).** *Let $G$ and $G'$ be two abelian groups and let $f\colon\ G \longrightarrow G'$ be a homomorphism. If $G' = A' + B'$ is a factorization and $A$ is a subset of $G$ such that $f\colon\ A \longrightarrow A'$ is a bijection, then $G = A + B$ is also a factorization, where $B = f^{-1}(B')$.*

**Proof.** First let us show that any element $\mathbf{g} \in G$ can be written as $\mathbf{a}+\mathbf{b}$, where $\mathbf{a} \in A$ and $\mathbf{b} \in B$. Assume that

$$f(\mathbf{g}) = \mathbf{a}' + \mathbf{b}',$$

where $\mathbf{a}' \in A'$ and $\mathbf{b}' \in B'$. We choose $\mathbf{a} \in A$ such that $f(\mathbf{a}) = \mathbf{a}'$. Then we get

$$f(\mathbf{g}-\mathbf{a}) = \mathbf{a}' + \mathbf{b}' - \mathbf{a}' = \mathbf{b}'.$$

Therefore, by taking $\mathbf{b} = \mathbf{g} - \mathbf{a}$, we get

$$\mathbf{g} = \mathbf{a} + \mathbf{b}.$$

Next let us establish the uniqueness. On the contrary, assume that

$$\mathbf{a}_1 + \mathbf{b}_1 = \mathbf{a}_2 + \mathbf{b}_2 \tag{8.7}$$

holds with $\mathbf{a}_i \in A$ and $\mathbf{b}_i \in B$. Then we get

$$f(\mathbf{a}_1) + f(\mathbf{a}_2) = f(\mathbf{a}_2) + f(\mathbf{b}_2).$$

Since $G' = A' + B'$ is a factorization, we get $f(\mathbf{a}_1) = f(\mathbf{a}_2)$ and therefore $\mathbf{a}_1 = \mathbf{a}_2$. Then it follows by (8.7) that $\mathbf{b}_1 = \mathbf{b}_2$. In other words, $G = A + B$ is a factorization. The lemma is proved. □

**Lemma 8.8 (Szabó, 1986).** *If there does exist a counterexample for Keller's conjecture, then there is a finite abelian group $G$, which is the direct sum of cyclic groups generated by $\mathbf{g}_1, \ldots, \mathbf{g}_m$ of order 4 and has a factorization*

$$G = H + [\mathbf{g}_1]_2 + \cdots + [\mathbf{g}_m]_2$$

*such that $2\mathbf{g}_i \notin H - H$ holds for all $i$.*

**Proof.** Assume that there is a counterexample for Keller's conjecture. By Lemma 8.6, there is a finite abelian group $G'$ with a factorization

$$G' = H' + [\mathbf{g}'_1]_2 + \cdots + [\mathbf{g}'_m]_2$$

such that $2\mathbf{g}'_i \notin H' - H'$.

Let $G^*$ denote the abelian group generated by $\mathbf{g}'_1, \ldots, \mathbf{g}'_m$ and write $H^* = H' \bigcap G^*$, then we have

$$G^* = H^* + [\mathbf{g}'_1]_2 + \cdots + [\mathbf{g}'_m]_2.$$

In $E^m$ let $\Lambda'$ denote the lattice generated by $\frac{1}{2}\mathbf{e}_1, \ldots, \frac{1}{2}\mathbf{e}_m$ and define

$$f\colon \sum \tfrac{1}{2} z_i \mathbf{e}_i \longrightarrow \sum z_i \mathbf{g}'_i.$$

It is easy to verify that $f$ is a homomorphism from $\Lambda'$ to $G^*$. Writing $X' = f^{-1}(H^*)$, by Lemma 8.7 we get

$$\Lambda' = \left[\tfrac{1}{2}\mathbf{e}_1\right]_2 + \cdots + \left[\tfrac{1}{2}\mathbf{e}_m\right]_2 + X'$$

and therefore $I^m + X'$ is a tiling of $E^m$. Next, we claim that $I^m + X'$ has no twin. Otherwise, if $\mathbf{e}_j \in X' - X'$, we can get

$$2\mathbf{g}'_j = f(\mathbf{e}_j) \in f(X') - f(X') = H^* - H^* \subset H' - H',$$

which contradicts the assumption.

Applying the method in the proof of Lemma 8.4 to $I^m + X'$, we can get that there is a counterexample $I^m + X$ for Keller's conjecture such that $X$ is a periodic set with periods $2\mathbf{e}_1, \ldots, 2\mathbf{e}_m$, $X \subset \Lambda'$, and

$$\Lambda' = \left[\tfrac{1}{2}\mathbf{e}_1\right]_2 + \cdots + \left[\tfrac{1}{2}\mathbf{e}_m\right]_2 + X. \tag{8.8}$$

Let $\Lambda$ denote the lattice generated by $2\mathbf{e}_1, \ldots, 2\mathbf{e}_m$ and write

$$H = \left\{\mathbf{x} \in X : \ 0 \le x_i < 2\right\},$$

it follows by (8.8) that

$$\Lambda' = H + \left[\tfrac{1}{2}\mathbf{e}_1\right]_2 + \cdots + \left[\tfrac{1}{2}\mathbf{e}_m\right]_2 + \Lambda.$$

Then the group

$$G = \Lambda'/\Lambda = H + [\mathbf{g}_1]_2 + \cdots + [\mathbf{g}_m]_2$$

satisfies all the requirements of the lemma, where $\mathbf{g}_i$ denote the elements corresponding to $\frac{1}{2}\mathbf{e}_i$. □

**Remark 8.1.** The number $m$ in Lemmas 8.6 and 8.8 is not directly related to the dimensions. However, this does not matter too much for our purpose, since Lemma 8.9 will guarantee the opposite.

**Lemma 8.9 (Szabó, 1986).** *Let $G$ be a direct sum of cyclic groups of order 4 generated by $\mathbf{g}_1, \ldots, \mathbf{g}_n$. If $G$ has a factorization*

$$G = H + [\mathbf{g}_1]_2 + \cdots + [\mathbf{g}_n]_2$$

*such that $2\mathbf{g}_i \notin H - H$ for all $i$, then there is a counterexample for Keller's conjecture in $E^n$.*

**Proof.** Let $\Lambda'$ be the lattice generated by $\{\frac{1}{2}\mathbf{e}_1, \ldots, \frac{1}{2}\mathbf{e}_n\}$ and define

$$f: \sum \tfrac{1}{2} z_i \mathbf{e}_i \longrightarrow \sum z_i \mathbf{g}_i.$$

It is easy to see that $f$ is a homomorphism from $\Lambda'$ on to $G$. Writing $X = f^{-1}(H)$, by Lemma 8.7 we have

$$\Lambda' = \left[\tfrac{1}{2}\mathbf{e}_1\right]_2 + \cdots + \left[\tfrac{1}{2}\mathbf{e}_n\right]_2 + X$$

and therefore $I^n + X$ is a tiling of $E^n$. If $\mathbf{e}_j \in X - X$, then we can get

$$2\mathbf{g}_j = f(\mathbf{e}_j) \in f(X) - f(X) = H - H,$$

which contradicts the assumption. Therefore $I^n + X$ has no twin. The lemma is proved. □

**Proof of the criterion.** It is well known that $G = A + B$ is a factorization of an abelian group $G$ if and only if

$$\{A - A\} \bigcap \{B - B\} = \{0\}$$

and

$$|G| = |A| \cdot |B|.$$

Writing

$$V = [\mathbf{g}_1]_2 + \cdots + [\mathbf{g}_n]_2,$$

the condition of Lemma 8.9 is equivalent to

$$|H| = 4^n/2^n = 2^n$$

and

$$\{H - H\} \bigcap (\{V - V\} \bigcup \{2\mathbf{g}_1, \ldots, 2\mathbf{g}_n\}) = \{0\}.$$

Define a graph $\Upsilon_n$ on $G$ that two elements $\mathbf{g}$ and $\mathbf{g}'$ are adjacent if and only if

$$\mathbf{g} - \mathbf{g}' \notin \{V - V\} \bigcup \{2\mathbf{g}_1, \ldots, 2\mathbf{g}_n\}.$$

Since

$$\{V - V\} = \{(z_1, \ldots, z_n) : \ z_i = 0, 1, 3\},$$

the graph $\Upsilon_n$ is isomorphic to the graph $\Gamma_n$ defined in Section 8.1. Let $\Psi_n$ denote one of the maximal clique of $\Upsilon_n$. Since every vertex in $\Upsilon_n$ plays the same role, we can assume that $\mathbf{0} \in \Psi_n$. Thus, if $\rho_n = 2^n$, by taking $H$ to be the vertices of $\Psi_n$, we get

$$G = H + [\mathbf{g}_1]_2 + \cdots + [\mathbf{g}_n]_2,$$

which satisfies the conditions of Lemma 8.9. Then we get a counterexample for Keller's conjecture. □

**Remark 8.2.** From the proof we notice two facts. First, $\rho_n$ cannot be larger than $2^n$. Second, there are many cliques of size $\rho_n$. Therefore, if $\rho_n = 2^n$, we do have a good chance to construct one of them.

## 8.4 Lagarias, Mackey, and Shor's counterexamples

In 1992, applying Corrádi and Szabó's criterion, J.C. Lagarias and P. Shor discovered the first counterexample for Keller's conjecture for $n \geq 10$. Ten years later, J. Mackey was able to reduce the dimensions to $n \geq 8$.

**Theorem 8.2 (Lagarias and Shor, 1992; Mackey, 2002).** *When $n \geq 8$, there exists a translative tiling $I^n + X$ of $E^n$, which has no twin.*

**Proof.** It is known that there will be such a tiling in $E^{n+1}$ if there is one in $E^n$. So it is sufficient to show the $n = 8$ case.

Let $\Psi_8$ be a subgraph of $\Gamma_8$ with the following 256 vertices.

(3, 1, 1, 1, 0, 2, 1, 1), (3, 1, 1, 1, 1, 1, 3, 2), (3, 1, 1, 1, 2, 3, 0, 3), (3, 1, 1, 1, 3, 0, 2, 0),
(3, 3, 2, 1, 0, 2, 1, 1), (3, 3, 2, 1, 1, 1, 3, 2), (3, 3, 2, 1, 2, 3, 0, 3), (3, 3, 2, 1, 3, 0, 2, 0),
(1, 0, 0, 3, 0, 2, 1, 1), (1, 0, 0, 3, 1, 1, 3, 2), (1, 0, 0, 3, 2, 3, 0, 3), (1, 0, 0, 3, 3, 0, 2, 0),
(1, 2, 3, 3, 0, 2, 1, 1), (1, 2, 3, 3, 1, 1, 3, 2), (1, 2, 3, 3, 2, 3, 0, 3), (1, 2, 3, 3, 3, 0, 2, 0),

(0, 0, 0, 0, 0, 0, 0, 0), (0, 0, 0, 0, 0, 2, 3, 0), (0, 0, 0, 0, 2, 1, 1, 2), (0, 0, 0, 0, 2, 3, 2, 2),
(0, 2, 3, 0, 0, 0, 0, 0), (0, 2, 3, 0, 0, 2, 3, 0), (0, 2, 3, 0, 2, 1, 1, 2), (0, 2, 3, 0, 2, 3, 2, 2),
(2, 1, 1, 2, 0, 0, 0, 0), (2, 1, 1, 2, 0, 2, 3, 0), (2, 1, 1, 2, 2, 1, 1, 2), (2, 1, 1, 2, 2, 3, 2, 2),
(2, 3, 2, 2, 0, 0, 0, 0), (2, 3, 2, 2, 0, 2, 3, 0), (2, 3, 2, 2, 2, 1, 1, 2), (2, 3, 2, 2, 2, 3, 2, 2),

(1, 0, 1, 1, 0, 2, 1, 1), (1, 0, 1, 1, 1, 1, 3, 2), (1, 0, 1, 1, 2, 3, 0, 3), (1, 0, 1, 1, 3, 0, 2, 0),
(1, 3, 3, 1, 0, 2, 1, 1), (1, 3, 3, 1, 1, 1, 3, 2), (1, 3, 3, 1, 2, 3, 0, 3), (1, 3, 3, 1, 3, 0, 2, 0),
(3, 1, 0, 3, 0, 2, 1, 1), (3, 1, 0, 3, 1, 1, 3, 2), (3, 1, 0, 3, 2, 3, 0, 3), (3, 1, 0, 3, 3, 0, 2, 0),
(3, 2, 2, 3, 0, 2, 1, 1), (3, 2, 2, 3, 1, 1, 3, 2), (3, 2, 2, 3, 2, 3, 0, 3), (3, 2, 2, 3, 3, 0, 2, 0),

(3, 2, 1, 0, 2, 2, 1, 1), (3, 2, 1, 0, 1, 1, 3, 0), (3, 2, 1, 0, 0, 3, 0, 3), (3, 2, 1, 0, 3, 0, 2, 2),
(1, 3, 0, 2, 2, 2, 1, 1), (1, 3, 0, 2, 1, 1, 3, 0), (1, 3, 0, 2, 0, 3, 0, 3), (1, 3, 0, 2, 3, 0, 2, 2),
(0, 0, 2, 1, 2, 2, 1, 1), (0, 0, 2, 1, 1, 1, 3, 0), (0, 0, 2, 1, 0, 3, 0, 3), (0, 0, 2, 1, 3, 0, 2, 2),
(2, 1, 3, 3, 2, 2, 1, 1), (2, 1, 3, 3, 1, 1, 3, 0), (2, 1, 3, 3, 0, 3, 0, 3), (2, 1, 3, 3, 3, 0, 2, 2),

(0, 1, 3, 1, 0, 2, 1, 1), (0, 1, 3, 1, 1, 1, 3, 2), (0, 1, 3, 1, 2, 3, 0, 3), (0, 1, 3, 1, 3, 0, 2, 0),
(2, 0, 2, 3, 0, 2, 1, 1), (2, 0, 2, 3, 1, 1, 3, 2), (2, 0, 2, 3, 2, 3, 0, 3), (2, 0, 2, 3, 3, 0, 2, 0),
(1, 2, 1, 2, 0, 2, 1, 1), (1, 2, 1, 2, 1, 1, 3, 2), (1, 2, 1, 2, 2, 3, 0, 3), (1, 2, 1, 2, 3, 0, 2, 0),
(3, 3, 0, 0, 0, 2, 1, 1), (3, 3, 0, 0, 1, 1, 3, 2), (3, 3, 0, 0, 2, 3, 0, 3), (3, 3, 0, 0, 3, 0, 2, 0),

(0, 1, 0, 2, 0, 0, 0, 0), (0, 1, 0, 2, 0, 2, 3, 0), (0, 1, 0, 2, 2, 1, 1, 2), (0, 1, 0, 2, 2, 3, 2, 2),
(0, 2, 2, 2, 0, 0, 0, 0), (0, 2, 2, 2, 0, 2, 3, 0), (0, 2, 2, 2, 2, 1, 1, 2), (0, 2, 2, 2, 2, 3, 2, 2),
(2, 0, 1, 0, 0, 0, 0, 0), (2, 0, 1, 0, 0, 2, 3, 0), (2, 0, 1, 0, 2, 1, 1, 2), (2, 0, 1, 0, 2, 3, 2, 2),
(2, 3, 3, 0, 0, 0, 0, 0), (2, 3, 3, 0, 0, 2, 3, 0), (2, 3, 3, 0, 2, 1, 1, 2), (2, 3, 3, 0, 2, 3, 2, 2),

(1, 2, 1, 0, 3, 1, 1, 1), (1, 2, 1, 0, 3, 3, 2, 1), (1, 2, 1, 0, 1, 0, 0, 3), (1, 2, 1, 0, 1, 2, 3, 3),
(3, 3, 0, 2, 3, 1, 1, 1), (3, 3, 0, 2, 3, 3, 2, 1), (3, 3, 0, 2, 1, 0, 0, 3), (3, 3, 0, 2, 1, 2, 3, 3),

(0, 0, 2, 3, 3, 1, 1, 1), (0, 0, 2, 3, 3, 3, 2, 1), (0, 0, 2, 3, 1, 0, 0, 3), (0, 0, 2, 3, 1, 2, 3, 3),
(2, 1, 3, 1, 3, 1, 1, 1), (2, 1, 3, 1, 3, 3, 2, 1), (2, 1, 3, 1, 1, 0, 0, 3), (2, 1, 3, 1, 1, 2, 3, 3),

(1, 2, 1, 0, 3, 0, 1, 3), (1, 2, 1, 0, 3, 3, 3, 3), (1, 2, 1, 0, 1, 1, 0, 1), (1, 2, 1, 0, 1, 2, 2, 1),
(3, 3, 0, 2, 3, 0, 1, 3), (3, 3, 0, 2, 3, 3, 3, 3), (3, 3, 0, 2, 1, 1, 0, 1), (3, 3, 0, 2, 1, 2, 2, 1),
(0, 0, 2, 3, 3, 0, 1, 3), (0, 0, 2, 3, 3, 3, 3, 3), (0, 0, 2, 3, 1, 1, 0, 1), (0, 0, 2, 3, 1, 2, 2, 1),
(2, 1, 3, 1, 3, 0, 1, 3), (2, 1, 3, 1, 3, 3, 3, 3), (2, 1, 3, 1, 1, 1, 0, 1), (2, 1, 3, 1, 1, 2, 2, 1),

(0, 1, 3, 3, 0, 2, 1, 3), (0, 1, 3, 3, 3, 1, 3, 2), (0, 1, 3, 3, 2, 3, 0, 1), (0, 1, 3, 3, 1, 0, 2, 0),
(2, 0, 2, 1, 0, 2, 1, 3), (2, 0, 2, 1, 3, 1, 3, 2), (2, 0, 2, 1, 2, 3, 0, 1), (2, 0, 2, 1, 1, 0, 2, 0),
(3, 2, 1, 2, 0, 2, 1, 3), (3, 2, 1, 2, 3, 1, 3, 2), (3, 2, 1, 2, 2, 3, 0, 1), (3, 2, 1, 2, 1, 0, 2, 0),
(1, 3, 0, 0, 0, 2, 1, 3), (1, 3, 0, 0, 3, 1, 3, 2), (1, 3, 0, 0, 2, 3, 0, 1), (1, 3, 0, 0, 1, 0, 2, 0),

(0, 0, 0, 0, 0, 0, 1, 2), (0, 0, 0, 0, 0, 3, 3, 2), (0, 0, 0, 0, 2, 1, 0, 0), (0, 0, 0, 0, 2, 2, 2, 0),
(0, 2, 3, 0, 0, 0, 1, 2), (0, 2, 3, 0, 0, 3, 3, 2), (0, 2, 3, 0, 2, 1, 0, 0), (0, 2, 3, 0, 2, 2, 2, 0),
(2, 1, 1, 2, 0, 0, 1, 2), (2, 1, 1, 2, 0, 3, 3, 2), (2, 1, 1, 2, 2, 1, 0, 0), (2, 1, 1, 2, 2, 2, 2, 0),
(2, 3, 2, 2, 0, 0, 1, 2), (2, 3, 2, 2, 0, 3, 3, 2), (2, 3, 2, 2, 2, 1, 0, 0), (2, 3, 2, 2, 2, 2, 2, 0),

(3, 2, 1, 0, 1, 0, 1, 1), (3, 2, 1, 0, 1, 3, 3, 1), (3, 2, 1, 0, 3, 1, 0, 3), (3, 2, 1, 0, 3, 2, 2, 3),
(1, 3, 0, 2, 1, 0, 1, 1), (1, 3, 0, 2, 1, 3, 3, 1), (1, 3, 0, 2, 3, 1, 0, 3), (1, 3, 0, 2, 3, 2, 2, 3),
(0, 0, 2, 1, 1, 0, 1, 1), (0, 0, 2, 1, 1, 3, 3, 1), (0, 0, 2, 1, 3, 1, 0, 3), (0, 0, 2, 1, 3, 2, 2, 3),
(2, 1, 3, 3, 1, 0, 1, 1), (2, 1, 3, 3, 1, 3, 3, 1), (2, 1, 3, 3, 3, 1, 0, 3), (2, 1, 3, 3, 3, 2, 2, 3),

(3, 2, 1, 0, 1, 1, 1, 3), (3, 2, 1, 0, 1, 3, 2, 3), (3, 2, 1, 0, 3, 0, 0, 1), (3, 2, 1, 0, 3, 2, 3, 1),
(1, 3, 0, 2, 1, 1, 1, 3), (1, 3, 0, 2, 1, 3, 2, 3), (1, 3, 0, 2, 3, 0, 0, 1), (1, 3, 0, 2, 3, 2, 3, 1),
(0, 0, 2, 1, 1, 1, 1, 3), (0, 0, 2, 1, 1, 3, 2, 3), (0, 0, 2, 1, 3, 0, 0, 1), (0, 0, 2, 1, 3, 2, 3, 1),
(2, 1, 3, 3, 1, 1, 1, 3), (2, 1, 3, 3, 1, 3, 2, 3), (2, 1, 3, 3, 3, 0, 0, 1), (2, 1, 3, 3, 3, 2, 3, 1),

(3, 0, 1, 3, 0, 2, 1, 3), (3, 0, 1, 3, 3, 1, 3, 2), (3, 0, 1, 3, 2, 3, 0, 1), (3, 0, 1, 3, 1, 0, 2, 0),
(3, 3, 3, 3, 0, 2, 1, 3), (3, 3, 3, 3, 3, 1, 3, 2), (3, 3, 3, 3, 2, 3, 0, 1), (3, 3, 3, 3, 1, 0, 2, 0),
(1, 1, 0, 1, 0, 2, 1, 3), (1, 1, 0, 1, 3, 1, 3, 2), (1, 1, 0, 1, 2, 3, 0, 1), (1, 1, 0, 1, 1, 0, 2, 0),
(1, 2, 2, 1, 0, 2, 1, 3), (1, 2, 2, 1, 3, 1, 3, 2), (1, 2, 2, 1, 2, 3, 0, 1), (1, 2, 2, 1, 1, 0, 2, 0),

(0, 1, 0, 2, 0, 0, 1, 2), (0, 1, 0, 2, 0, 3, 3, 2), (0, 1, 0, 2, 2, 1, 0, 0), (0, 1, 0, 2, 2, 2, 2, 0),
(0, 2, 2, 2, 0, 0, 1, 2), (0, 2, 2, 2, 0, 3, 3, 2), (0, 2, 2, 2, 2, 1, 0, 0), (0, 2, 2, 2, 2, 2, 2, 0),
(2, 0, 1, 0, 0, 0, 1, 2), (2, 0, 1, 0, 0, 3, 3, 2), (2, 0, 1, 0, 2, 1, 0, 0), (2, 0, 1, 0, 2, 2, 2, 0),
(2, 3, 3, 0, 0, 0, 1, 2), (2, 3, 3, 0, 0, 3, 3, 2), (2, 3, 3, 0, 2, 1, 0, 0), (2, 3, 3, 0, 2, 2, 2, 0),

(1, 1, 1, 3, 0, 2, 1, 3), (1, 1, 1, 3, 3, 1, 3, 2), (1, 1, 1, 3, 2, 3, 0, 1), (1, 1, 1, 3, 1, 0, 2, 0),
(1, 3, 2, 3, 0, 2, 1, 3), (1, 3, 2, 3, 3, 1, 3, 2), (1, 3, 2, 3, 2, 3, 0, 1), (1, 3, 2, 3, 1, 0, 2, 0),
(3, 0, 0, 1, 0, 2, 1, 3), (3, 0, 0, 1, 3, 1, 3, 2), (3, 0, 0, 1, 2, 3, 0, 1), (3, 0, 0, 1, 1, 0, 2, 0),
(3, 2, 3, 1, 0, 2, 1, 3), (3, 2, 3, 1, 3, 1, 3, 2), (3, 2, 3, 1, 2, 3, 0, 1), (3, 2, 3, 1, 1, 0, 2, 0),

(1, 2, 1, 0, 2, 2, 1, 3), (1, 2, 1, 0, 3, 1, 3, 0), (1, 2, 1, 0, 0, 3, 0, 1), (1, 2, 1, 0, 1, 0, 2, 2),
(3, 3, 0, 2, 2, 2, 1, 3), (3, 3, 0, 2, 3, 1, 3, 0), (3, 3, 0, 2, 0, 3, 0, 1), (3, 3, 0, 2, 1, 0, 2, 2),
(0, 0, 2, 3, 2, 2, 1, 3), (0, 0, 2, 3, 3, 1, 3, 0), (0, 0, 2, 3, 0, 3, 0, 1), (0, 0, 2, 3, 1, 0, 2, 2),
(2, 1, 3, 1, 2, 2, 1, 3), (2, 1, 3, 1, 3, 1, 3, 0), (2, 1, 3, 1, 0, 3, 0, 1), (2, 1, 3, 1, 1, 0, 2, 2).

It can be verified that $\Psi_8$ is a clique of size 256 in $\Gamma_8$ and hence

$$\rho_8 = 256 = 2^8.$$

Then it follows by Corrádi and Szabó's criterion that Keller's conjecture is false in $E^8$.

In fact, this clique can be generated by 16 Cartesian products

$$(\mathbf{a},\mathbf{b}),\quad (\mathbf{c},\mathbf{c}),\quad (\mathbf{c},\mathbf{b}),\quad (\mathbf{e},\mathbf{f}),\quad (\mathbf{g},\mathbf{b}),\quad (\mathbf{h},\mathbf{c}),\quad (\mathbf{i},\mathbf{a}),\quad (\mathbf{i},\mathbf{j}),$$
$$(\mathbf{k},\mathbf{l}),\quad (\mathbf{c},\mathbf{m}),\quad (\mathbf{e},\mathbf{d}),\quad (\mathbf{e},\mathbf{n}),\quad (\mathbf{j},\mathbf{l}),\quad (\mathbf{h},\mathbf{m}),\quad (\mathbf{n},\mathbf{l}),\quad (\mathbf{i},\mathbf{o}),$$

where

$$\begin{aligned}
\mathbf{a} &= (3,1,1,1),\ (3,3,2,1),\ (1,0,0,3),\ (1,2,3,3),\\
\mathbf{b} &= (0,2,1,1),\ (1,1,3,2),\ (2,3,0,3),\ (3,0,2,0),\\
\mathbf{c} &= (0,0,0,0),\ (0,2,3,0),\ (2,1,1,2),\ (2,3,2,2),\\
\mathbf{d} &= (1,0,1,1),\ (1,3,3,1),\ (3,1,0,3),\ (3,2,2,3),\\
\mathbf{e} &= (3,2,1,0),\ (1,3,0,2),\ (0,0,2,1),\ (2,1,3,3),\\
\mathbf{f} &= (2,2,1,1),\ (1,1,3,0),\ (0,3,0,3),\ (3,0,2,2),\\
\mathbf{g} &= (0,1,3,1),\ (2,0,2,3),\ (1,2,1,2),\ (3,3,0,0),\\
\mathbf{h} &= (0,1,0,2),\ (0,2,2,2),\ (2,0,1,0),\ (2,3,3,0),\\
\mathbf{i} &= (1,2,1,0),\ (3,3,0,2),\ (0,0,2,3),\ (2,1,3,1),\\
\mathbf{j} &= (3,0,1,3),\ (3,3,3,3),\ (1,1,0,1),\ (1,2,2,1),\\
\mathbf{k} &= (0,1,3,3),\ (2,0,2,1),\ (3,2,1,2),\ (1,3,0,0),\\
\mathbf{l} &= (0,2,1,3),\ (3,1,3,2),\ (2,3,0,1),\ (1,0,2,0),\\
\mathbf{m} &= (0,0,1,2),\ (0,3,3,2),\ (2,1,0,0),\ (2,2,2,0),\\
\mathbf{n} &= (1,1,1,3),\ (1,3,2,3),\ (3,0,0,1),\ (3,2,3,1),\\
\mathbf{o} &= (2,2,1,3),\ (3,1,3,0),\ (0,3,0,1),\ (1,0,2,2).
\end{aligned}$$

The theorem is proved. □

**Remark 8.3.** It was proved by Robinson (1979) that every multiple tiling $I^2 + X$ of $E^2$ has a twin. On the other hand, Szabó (1982) proved that, whenever $n \geq 3$ and $k \geq 2$, there is a $k$-fold translative tiling $I^n + X$ of $E^n$ which has no twin.

# References


Ábrego, B. M., Fernández–Merchant, S., Neubauer, M. G., and Watkins, W. (2003). D-optimal weighing designs for n $\equiv -1$ (mod 4) objects and a large number of weighing, *Linear Algebra Appl.* **374**: 175–218.

Affentranger, F. and Schneider, R. (1992). Random projections of regular simplices, *Discrete Comput. Geom.* **7**: 219–226.

Agaian, S. S. (1985). *Hadamard Matrices and Their Applications*, Berlin: Springer-Verlag.

Aichholzer, O. (2000). Extremal properties of 0/1 polytopes of dimension 5, *Polytopes–Combinatorics and Computation, DMV Sem.* **29**: 111–130.

Aigner, M. and Ziegler, G. M. (2004). *Proofs from the Book* (third edition), Berlin: Springer-Verlag.

Ball, K. (1986). Cube slicing in $R^n$, *Proc. Amer. Math. Soc.* **97**: 465–473.

(1988). Logarithmically concave functions and sections of convex sets in $R^n$, *Studia Math.* **88**: 69–84.

(1989). Volumes of sections of cubes and related problems, *Lecture Notes in Math.* **1376**: 251–260.

(2001). Convex geometry and functional analysis, *Handbook of the Geometry of Banach Spaces*, Amsterdam: North-Holland, Vol. I, pp.161–194.

Bárány, I. and Lovász, L. (1982). Borsuk's theorem and the number of facets of centrally symmetric polytopes, *Acta Math. Acad. Sci. Hungar.* **40**: 323–329.

Bárány, I. and Pór, A. (2001). On $0-1$ polytopes with many facets, *Adv. Math.* **161**: 209–228.

Barba, G. (1933). Intorno al teorema di Hadamard sui determinanti a valore massimo, *Giorn. Mat. Battaglia* **71**: 70–86.

Barthe, F. and Koldobsky, A. (2003). Extremal slabs in the cube and the Laplace transform, *Adv. Math.* **174**: 89–114.

Beckner, W. (1975). Inequalities in Fourier analysis, *Ann. Math.* **102**: 159–182.

Below, A., Brehm, U., De Loera, J. A., and Richter-Gebert, J. (2000). Minimal simplicial dissections and triangulations of convex 3-polytopes, *Discrete Comput. Geom.* **24**: 35–48.

Billera, L. J., Cushman, R., and Sanders, J. A. (1988). The Stanley decomposition of the harmonic oscillator, *Nederl. Akad. Wetensch. Indag. Math.* **50**: 375–393.

Billera, L. J. and Sarangarajan, A. (1996). All $0-1$ polytopes are traveling salesman polytopes, *Combinatorica* **16**: 175–188.

Birnbaum, Z. W. (1948). On random variables with comparable peakedness, *Ann. Math. Stat.* **19**: 76–81.

Bliss, A. and Su, F. E. (2005). Lower bounds for simplicial covers and triangulations of cubes, *Discrete Comput. Geom.* **33**: 669–686.

Bonnesen, T. and Fenchel, W. (1934). *Theorie der konvexen Körper*, Berlin: Springer-Verlag.

Borell, C. (1975). Convex set functions in $d$-space, *Period. Math. Hungar.* **6**: 111–136.

Böröczky, Jr. K. and Henk, M. (1999). Random projections of regular polytopes, *Arch. Math.* **73**: 465–473.

Brascamp, H. J. and Lieb, E. H. (1976). Best constants in Young's inequality, its converse, and its generalization to more than three functions, *Adv. Math.* **20**: 151–173.

Brascamp, H. J., Lieb, E. H., and Luttinger, J. M. (1974). A general rearrangement inequality for multiple integrals, *J. Funct. Anal.* **17**: 227–237.

Brenner, J. and Cummings, L. (1972). The Hadamard maximum determinant problem, *Amer. Math. Monthly* **79**: 626–629.

Broadie, M. N. and Cottle, R. W. (1984). A note on triangulating the 5-cube, *Discrete Math.* **52**: 39–49.

Brualdi, R. and Solheid, E. S. (1986). Maximum determinants of complementary acyclic matrices of zeros and ones, *Discrete Math.* **61**: 1–19.

Chakerian, G. D. and Filliman, P. (1986). The measures of the projections of a cube, *Studia Sci. Math. Hungar.* **21**: 103–110.

Christof, T. and Reinelt, G. (2001). Decomposition and parallelization techniques for enumerating the facets for combinatorial polytopes, *J. Comput. Geom. Appl.* **11**: 423–437

Clements, G. F. and Lindström, B. (1965). A sequence of ($\pm 1$)-determinants with large values, *Proc. Amer. Math. Soc.* **16**: 548–550.

Corrádi, K. and Szabó, S. (1990a). Cube tiling and covering a complete graph, *Discrete Math.* **85**: 319–321.

(1990b). A combinatorial approach for Keller's conjecture, *Period. Math. Hungar.* **21**: 95–100.

Cottle, R. W. (1982). Minimal triangulations of the 4-cube, *Discrete Math.* **40**: 25–29.

Delone, B. N. (1929). Sur la partition reguliére de l'espace à 4 dimensions I, II, *Izv. Akad. Nauk SSSR* **7**: 79–110, 147–164.

Delsarte, P. (1972). Bounds for unrestricted codes, by linear programming, *Philips Research Reports* **27**: 272–289.

(1973). An algebraic approach to the association schemes of coding theory, *Philips Research Reports, Suppl.* **10**: 1–97.

Derry, D. (1940). Remarks on a conjecture of Minkowski, *Amer. J. Math.* **62**: 61–66.

Dvoretzky, A. (1961). Some results on convex bodies and Banach spaces, *Proc. Internat. Symp. Linear Spaces*, 123–160.

(1963). Some near-sphericity results, *Proc. Symp. Pure Math.* **7**: 203–210.

Dyer, M. E., Füredi, Z., and McDiarmid, C. (1992). Volumes spanned by random points in the hypercube, *Random Structures Algorithms* **3**: 91–106.

Ehlich, H. (1964a). Determinantenabschätzungen für binäre Matrizen, *Math. Z.* **83**: 123–132.

(1964b). Determinantenabschätzungen für binäre Matrizen mit $n \equiv 3 \bmod 4$, *Math. Z.* **84**: 438–447.

Ehlich, H. and Zeller, K. (1962). Binäre Matrizen, *Z. Angew. Math. Mech*, **42**: 20–21.

Engel, P. (1993). Geometric crystallography, *Handbook of Convex Geometry* (P. M. Gruber and J. M. Wills eds), Amsterdam: North-Holland, pp.989–1042.

Erdös, P., Gruber, P. M., and Hammer, J. (1989). *Lattice points*, New York: Longman Scientific & Technical.

Fedorov, E. S. (1885). *Elements of the theory of figures*, St. Petersburg: Imp. Acad. Sci..

Fejes Tóth, L. (1964). *Regular Figures,* London: Pergamon Press.

Fleiner, T., Kaibel, V., and Rote, G. (2000). Upper bounds on the maximal number of facets of 0/1 polytopes, *European J. Combin.* **21**: 121–130.

Fuchs, L. (1958). *Abelian Groups*, Budapest: Publishing House of the Hungarian Academy of Sciences.

Furtwängler, Ph. (1936). Über Gitter konstanter Dichte, *Monatsh. Math. Phys.* **43**: 281–288.

Gardner, R. J. (1995). *Geometric Tomography*, Cambridge: Cambridge University Press.

Gatzouras, D., Giannopoulos, A., and Markoulakis, N. (2005). Lower bound for the maximal number of facets of a 0/1 polytope, *Discrete Comput. Geom.*, in press.

Gilbert, E. N. (1952). A comparison of signaling alphabets, *Bell System Tech. J.* **31**: 504–522.

Grigorév, N. A. (1982). Regular simplices inscribed in a cube and Hadamard matrices, *Proc. Steklov Inst. Math.* **152**: 97–98.

Gritzmann, P., Klee, V., and Larman, D. G. (1995). Largest $j$-simplices in $n$-polytopes, *Discrete Comput. Geom.* **13**: 477–515.

Grünbaum, B. (1967). *Convex Polytopes,* London: Intersciences.

(1968). Grassmann angles of convex polytopes, *Acta Math.* **121**: 293–302.

Gruner, W. (1939/40). Einlagerung des regulären $n$-Simplex in den $n$-dimensionalen Würfel, *Comment. Math. Helvetici* **12**: 149–152.

Haagerup, U. and Munkholm, H. J. (1981). Simplices of maximal volume in hyperbolic $n$-space, *Acta Math.* **147**: 1–12.

Hadamard, J. (1893). Résolution d'une question relativ aux déterminants, *Bull. Sci. Math.* **28**: 240–246.

Hadwiger, H. (1972). Gitterperiodische Punktmengen und Isoperimetrie, *Monatsh. Math.* **76**: 410–418.

Haiman, M. (1991). A simple and relatively efficient triangulation of the $n$-cube, *Discrete Comput. Geom.* **6**: 287–289.

Hajós, G. (1938). Covering high dimensional spaces by cube lattices, *Mat. Fiz. Lapok* **45**: 171–190.

(1941). Über einfache und mehrfache Bedeckung des $n$-dimensionalen Raumes mit einem Würfelgitter, *Math. Z.* **47**: 427–467.

(1949). Sur la factorisation des groupes abéliens, *Časopis Pěst. Mat. Fys.* **74**: 157–162.

Hall, Jr. M. (1966). *Combinatorial Theory*, Waltham, MA: Blaisdell: 1966.

Hensley, D. (1979). Slicing the cube in $R^n$ and probability, *Proc. Amer. Math. Soc.* **73**: 95–100.
(1980). Slicing convex bodies–bounds for slice area in terms of the body's covariance, *Proc. Amer. Math. Soc.* **79**: 619–625.
Hudelson, M., Klee, V., and Larman, D. G. (1996). Largest $j$-simplices in $d$-cubes: Some relatives of the Hadamard maximum determinant problem, *Linear Algebra Appl.* **241–243**: 519–598.
Hughes, R. B. (1993). Minimum-cardinality triangulations of the $d$-cube for $d = 5$ and $d = 6$, *Discrete Math.* **118**: 75–118.
(1994). Lower bounds on cube simplexity, *Discrete Math.* **133**: 123–138.
Hughes, R. B. and Anderson, M. R. (1993). A triangulation of the 6-cube with 308 simplices, *Discrete Math.* **117**: 253–256.
(1996). Simplexity of the cube, *Discrete Math.* **158**: 99–150.
Jaffe, D. B. (2000). Optimal binary linear codes of length $\leq 30$, *Discrete Math.* **223**: 135–155.
Kahn, J., Komlós, J., and Szemerédi, E. (1995). On the probability that a random $\pm 1$-matrix is singular, *J. Amer. Math. Soc.* **8**: 223–240.
Kanter, M. (1977). Unimodality and dominance for symmetric random vectors, *Trans. Amer. Math. Soc.* **229**: 65–85.
Keller, O. H. (1930). Über die lückenlose Erfüllung des Raumes mit Würfeln, *J. reine angew. Math.* **163**: 231–248.
(1937). Ein Satz über die lückenlose Erfüllung des 5- und 6-dimensionalen Raumes mit Würfeln, *J. reine angew. Math.* **177**: 61–64.
Kolountzakis, M. (1998). Lattice tilings by cubes: whole, notched and extended, *Electron. J. Combin.* **5**: 1–11.
Kortenkamp, U. H., Richter-Gebert, J., Sarangarajan, A., and Ziegler, G. M. (1997). Extremal properties of 0/1-polytopes, *Discrete Comput. Geom.* **17**: 439–448.
Kowalewski, G. (1954). *Einführung in der Determinantentheorie einschliesslich der Fredholmschen Determinanten*, Berlin: Walter de Gruyter & Co.
Lagarias, J. C. and Shor, P. (1992). Keller's cube-tiling conjecture is false in high dimensions, *Bull. Amer. Math. Soc.* **27**: 279–283.
(1994). Cube tilings and nonlinear codes, *Discrete Comput. Geom.* **11**: 359–391.
Larman, D. G. and Mani, P. (1975). Almost ellipsoidal sections and projections of convex bodies, *Proc. Cambridge Philos. Soc.* **77**: 529–546.
Lee, C. (1985). Triangulating the $d$-cube, *Discrete Geometry and Convexity* (J. E. Goodman, E. Lutwak, J. Malkevitch, and R. Pollack, eds.), New York: New York Academy of Sciences, pp.205–211.
(1997). Subdivisions and triangulations of polytopes, *Handbook of Discrete and Computational Geometry*, Boca Raton: CRC Press, pp.271–290.
Lonke, Y. (2000). On random sections of the cube, *Discrete Comput. Geom.* **23**: 157–169.
Mackey, J. (2002). A cube tiling of dimension eight with no facesharing, *Discrete Comput. Geom.* **28**: 275–279.
Mara, P. S. (1976). Triangulations for the cube, *J. Combin. Theory* (*A*), **20**: 170–177.
McEliece, R. J., Rodemich, E. R., Rumsey Jr. H., and Welch, L. R. (1977). New upper bounds on the rate of a code via the Delsarte-Macwilliams inequalities, *IEEE Trans. Inform. Theory* **23**: 157–166.

McMullen, P. (1975). Non-linear angle-sum relations for polyhedral cones and polytopes, *Math. Proc. Cambridge Philos. Soc.* **78**: 247–261.

(1984). Volumes of projections of unit cubes, *Bull. London Math. Soc.* **16**: 278–280.

Medyanik, A. I. (1997). A regular simplex inscribed in a cube and a Hadamard matrix of half circulant type, *Mat. Fiz. Anal. Geom.* **4**: 458–471.

(2001). A regular simplex inscribed in a cube, half circulant Hadamard matrices, and Gaussian sums, *Mat. Fiz. Anal. Geom.* **8**: 58–81.

Meyer, M. and Pajor, A. (1988). Section of the unit ball of $\ell_p^n$, *J. Funct. Anal.* **80**: 109–123.

Milnor, J. (1982). Hyperbolic geometry – the first 150 years, *Bull. Amer. Math. Soc.* **6**: 9–24.

Minkowski, H. (1896). *Geometrie der Zahlen*, Leipzig: Teubner.

(1907). *Diophantische Approximationen*, Leipzig: Teubner.

Miyamoto, M. (1991). A construction of Hadamard matrices, *J. Combin. Theory* (*A*) **57**: 86–108.

Mordell, L. J. (1936). Minkowski's theorems and hypotheses on linear forms, *Proc. ICM*, Oslo, pp.226–238.

Neubauer, M. G. and Radcliffe, A. J. (1997). The maximum determinant of $\pm 1$ matrices, *Linear Algebra Appl.* **257**: 289–306.

Neubauer, M. G. and Watkins, W. (2002). D-optimal designs for seven objects and a large number of weighings, *Linear Multilinear Algebra* **50**: 61–74.

Neubauer, M. G., Watkins, W., and Zeitlin, J. (1997). Maximal $j$-simplices in the real $d$-dimensional unit cube, *J. Combin. Theory* (*A*) **80**: 1–12.

(1998a). Notes on D-optimal designs *Linear Algebra Appl.* **280**: 109–127.

(1998b). D-optimal weighing designs for four and five objects, *Electron. J. Linear Algebra* **4**: 48–73.

(2000). D-optimal weighing designs for six objects, *Metrika* **52**: 185–211.

Orden, D. and Santos, F. (2003). Asymptotically efficient triangulations of the $d$-cube, *Discrete Comput. Geom.* **30**: 509–528.

Ostrovskii, M. I. (2000). Minimal volume shadows of cubes, *J. Funct. Anal.* **176**: 317–330.

Paley, R. E. A. C. (1933). On orthogonal matrices, *J. Math. Phys.* **12**: 311–320.

Perron, O. (1940a). Über lückenlose Ausfüllung des $n$-dimensionalen Raumes durch kongruente Würfel I; II, *Math. Z.* **46**: 1–26, 161–180.

(1940b). Modulartige lückenlose Ausfüllung des $R^n$ mit kongruenten Würfeln I; II, *Math. Ann.* **117**: 415–447; **117** (1941): 609–658.

Pless, V. S., Huffman, W. C., and Brualdi, R. A. (eds) (1998). *Handbook of Coding Theory*, Elsevier.

Prékopa, A. (1971). Logarithmic concave measures with applications to stochastic programming, *Acta Sci. Math.* **32**: 301–306.

(1973). On logarithmic concave measures and functions, *Acta Sci. Math.* (Szeged) **34**: 335–343.

Raktoe, L. (1979). How many degenerate simplices are generated by $n+1$ vertices of the unit $n$-cube, *Amer. Math. Monthly* **86**: 49.

Robinson, R. M. (1979). Multiple tilings of $n$-dimensional space by unit cubes, *Math. Z.* **166**: 225–275.

Sallee, J. F. (1982a). A triangulation of the $n$-cube, *Discrete Math.* **40**: 81–86.

(1982b). A note on minimal triangulations of an $n$-cube, *Discrete Applied Math.* **4**: 211–215.

(1984). Middle-cut triangulations of the $n$-cube, *SIAM J. Algbraic Discrete Methods* **5**: 407–419.

Santaló, L. A. (1952). Geometria integral en espacios de curvatura constante, *Rep. Argentina Publ. Com. Nac. Energia Atómica, Ser. Mat.* **1**.

Sawade, K. (1985). A Hadamard matrix of order 268, *Graphs Combin.* **1**: 185–187.

Schmidt, K. W. (1970). Lower bounds for maximal (0, 1)-determinants, *SIAM J. Appl. Math.* **19**: 440–441.

Schmidt, T. (1933). Über die Zerlegung des $n$-dimensionalen Raumes in gitterförmig angeordnete Würfeln, *Schr. Math. Semin. U. Institut angew. Math. Univ. Berlin* **1**: 186–212.

Schneider, R. (1993). *Convex Bodies: The Brunn-Minkowski Theory*, Cambridge: Cambridge University Press.

Schulte, E. (1993). Tilings, *Handbook of Convex Geometry* (P. M. Gruber and J. M. Wills eds.), Amsterdam: North-Holland, pp.899–932.

Seberry, J. R. and Yamada, M. (1992). Hadamard matrices, sequences, and block designs, *Contemporary Design Theory – A Collection of Surveys* (J. H. Dinitz and D. R. Stinson eds.), New York: Wiley, pp.431–560.

Shephard, G. C. (1974). Combinatorial properties of associated zonotopes, *Canad. J. Math.* **26**: 302–321.

Sloane, N. J. A. (1977). Error-correcting codes and invariant theory: new applications of a nineteenth-century technique, *Amer. Math. Monthly* **84**: 82–107.

Smith, W. D. (2000). A lower bound for the simplexity of the $n$-cube via hyperbolic volumes, *European J. Combin.* **21**: 131–137.

Stein, S. K. (1972). A symmetric star body that tiles but not as a lattice, *Proc. Amer. Math. Soc.* **36**: 543–548.

(1974). Algebraic tiling, *Amer. Math. Monthly* **81**: 445–462.

Stein, S. K. and Szabó, S. (1994). *Algebra and Tiling: Homomorphisms in the Service of Geometry*, Washington DC: Math. Assoc. Amer.

Štogrin, M. I. (1975). Regular Dirichlet-Voronoi partitions for the second triclinic group, *Proc. Steklov Inst. Math.* **123**.

Sylvester, J. J. (1867). Thoughts on inverse orthogonal matrices, simultaneous sign-succession, and tesselated pavements in two or more colours, with applications to Newton's rule, ornamental tile-work, and the theory of numbers, *London, Edingburgh and Dublin Philos. Mag. and J. Sci.* **34**: 461–475.

Szabó, S. (1982). Multiple tilings by cubes with no shared faces, *Aequationes Math.* **25**: 83–89.

(1986). A reduction of Keller's conjecture, *Period. Math. Hungar.* **17**: 265–277.

(1993). Cube tilings as contributions of algebra to geometry, *Beiträge Algebra Geom.* **34**: 63–75.

Thurston, W. P. (1977). *The Geometry and Topology of 3-manifolds*, Lecture Notes from Princeton University.

Vaaler, J. D. (1979). A geometric inequality with applications to linear forms, *Pacific J. Math.* **83**: 543–553.

van Lint, J. H. (1982). *Introduction to Coding Theory*, New York: Springer-Verlag.

Varshamov, R. R. (1957). Estimate of the number of signals in error correcting codes, *Dokl. Akad. Nauk SSSR* **117**: 739–741.

Voronoi, G. F. (1908). Nouvelles applications des paramètres continus à la théorie des formes quadratiques, *J. reine angew. Math.* **134**: 198–287; **136** (1909): 67–181.

Williamson, J. (1946). Determinants whose elements are 0 and 1, *Amer. Math. Monthly* **53**: 427–434.

Wojtas, M. (1964). On Hadamard's inequality for the determinants of order non-divisible by 4, *Colloq. Math.* **12**: 73–83.

Yamada, M. (1989). Some new series of Hadamard matrices, *J. Austral. Math. Soc.* **46**: 371–383.

Yang, C. H. (1966a). Some designs for maximal $(+1, -1)$ determinant of order $n \equiv 2$ (mod 4), *Math. Comp.* **20**: 147–148; **22** (1968): 174–180; **23** (1969): 201–205.

(1966b). A construction for maximal $(+1, -1)$ matrix of order 54, *Bull. Amer. Math. Soc.* **72**: 293.

Ziegler, G. M. (2000). Lectures on 0/1-Polytopes, *Polytopes – Combinatorics and Computation, DMV Sem.* **29**: 1–41.

Zong, C. (1996). *Strange Phenomena in Convex and Discrete Geometry*, New York: Springer-Verlag.

(2005). What is known about unit cubes, *Bull. Amer. Math. Soc.* **42**: 181–211.

# Index